DATE DUE

VOLUME EIGHTY THREE

Advances in
GENETICS

ADVANCES IN GENETICS, VOLUME 83

Serial Editors

Theodore Friedmann
University of California at San Diego,
School of Medicine, CA, USA

Jay C. Dunlap
The Geisel School of Medicine at Dartmouth,
Hanover, NH, USA

Stephen F. Goodwin
University of Oxford, Oxford, UK

> VOLUME EIGHTY THREE

ADVANCES IN
GENETICS

Edited by

Theodore Friedmann
Department of Pediatrics,
University of California at San Diego,
School of Medicine, CA, USA

Jay C. Dunlap
Department of Genetics,
The Geisel School of Medicine at Dartmouth,
Hanover, NH, USA

Stephen F. Goodwin
Department of Physiology,
Anatomy and Genetics,
University of Oxford,
Oxford, UK

AMSTERDAM · BOSTON · HEIDELBERG · LONDON
NEW YORK · OXFORD · PARIS · SAN DIEGO
SAN FRANCISCO · SINGAPORE · SYDNEY · TOKYO

Academic Press is an imprint of Elsevier

ELSEVIER

Academic Press is an imprint of Elsevier
225 Wyman Street, Waltham, MA 02451, USA
525 B Street, Suite 1800, San Diego, CA 92101-4495, USA
Radarweg 29, PO Box 211, 1000 AE Amsterdam, The Netherlands
The Boulevard, Langford Lane, Kidlington, Oxford, OX5 1GB, UK
32 Jamestown Road, London, NW1 7BY, UK

First edition 2013

ISBN: 978-0-12-407675-4
ISSN: 0065-2660

For information on all Academic Press publications
visit our website at store.elsevier.com

Printed and bound in USA
13 14 15   10 9 8 7 6 5 4 3 2 1

Working together
to grow libraries in
developing countries

www.elsevier.com • www.bookaid.org

CONTENTS

Contributors   *vii*

1. Biology and Mechanisms of Short RNAs in *Caenorhabditis elegans*   1
Alla Grishok

   1. Introduction   2
   2. miRNA Function   3
   3. RNAi and Endogenous siRNAs   13
   4. 21U-RNAs, *C. elegans* piRNAs   47
   5. Systemic Features of RNAi   54
   6. Outlook   57
   Acknowledgments   57
   References   58

2. Genetics of Immune Recognition and Response in *Drosophila* host defense   71
Petros Ligoxygakis

   1. Introduction   72
   2. Toll Signaling in *Drosophila*   75
   3. The Immune Deficiency Pathway   83
   4. Immunity as a Sensor of Metabolism   87
   5. Concluding Remarks—Future Directions   88
   Acknowledgments   89
   References   89

3. Recent Advances in Septum Biogenesis in *Neurospora crassa*   99
Rosa Reyna Mouriño-Pérez and Meritxell Riquelme

   1. Introduction   100
   2. Septum Ontogeny   101
   3. Septum Wall Biosynthesis   113
   4. Septal Pores and Associated Proteins   119
   5. Conclusions   120
   6. Outlook and Future Research   122
   Acknowledgments   123
   References   123

Index   *135*

CONTRIBUTORS

Alla Grishok
Department of Biochemistry and Molecular Biophysics, Columbia University, New York, New York, USA

Petros Ligoxygakis
Laboratory of Genes and Development, Department of Biochemistry, University of Oxford, Oxford, United Kingdom

Rosa Reyna Mouriño-Pérez
Departamento de Microbiología, Centro de Investigación Científica y de Educación Superior de Ensenada, CICESE, Ensenada, Baja California, Mexico

Meritxell Riquelme
Departamento de Microbiología, Centro de Investigación Científica y de Educación Superior de Ensenada, CICESE, Ensenada, Baja California, Mexico

CHAPTER ONE

Biology and Mechanisms of Short RNAs in *Caenorhabditis elegans*

Alla Grishok[1]

Department of Biochemistry and Molecular Biophysics, Columbia University, New York, New York, USA
[1]Corresponding author: e-mail address: ag2691@columbia.edu

Contents

1. Introduction	2
2. miRNA Function	3
2.1 Biological effects of miRNAs	3
2.2 Biogenesis and molecular mechanisms of miRNAs	9
3. RNAi and Endogenous siRNAs	13
3.1 Exogenous RNAi pathway	14
3.2 Genomic and molecular features of endogenous RNAi	29
3.3 Biological functions of RNAi	37
4. 21U-RNAs, *C. elegans* piRNAs	47
4.1 Biogenesis of 21U-RNAs	47
4.2 Biological functions of 21U-RNAs	51
5. Systemic Features of RNAi	54
5.1 dsRNA import channel SID-1 and the features of mobile RNA species	54
5.2 Additional factors required for systemic and environmental RNAi	56
6. Outlook	57
Acknowledgments	57
References	58

Abstract

The significance of noncoding RNAs in animal biology is being increasingly recognized. The nematode *Caenorhabditis elegans* has an extensive system of short RNAs that includes microRNAs, piRNAs, and endogenous siRNAs, which regulate development, control life span, provide resistance to viruses and transposons, and monitor gene duplications. Progress in our understanding of short RNAs was stimulated by the discovery of RNA interference, a phenomenon of sequence-specific gene silencing induced by exogenous double-stranded RNA, at the turn of the twenty-first century. This chapter provides a broad overview of the exogenous and endogenous RNAi processes in *C. elegans* and describes recent advances in genetic, genomic, and molecular analyses of nematode's short RNAs and proteins involved in the RNAi-related pathways.

Advances in Genetics, Volume 83
ISSN 0065-2660
http://dx.doi.org/10.1016/B978-0-12-407675-4.00001-8

1. INTRODUCTION

Within the last 20 years, it has been recognized that short RNAs 20–30 nt in length and their protein cofactors of the ancient Argonaute family are integral components of most eukaryotic organisms. Some short RNAs, such as the microRNA *let-7*, are well conserved in animals (Pasquinelli et al., 2000), while others, such as endogenous siRNAs antisense to foreign RNA sequences and involved in genome surveillance, can be generated within the lifetime of an individual (reviewed in Billi, Freeberg, & Kim, 2012).

The nematode *Caenorhabditis elegans* represents an ideal organism for studies of short RNA biology. The first microRNAs to be discovered, *lin-4* (Lee, Feinbaum, & Ambros, 1993) and *let-7* (Reinhart et al., 2000), were found in the course of rigorous genetic analysis of the *C. elegans* cell lineage, and the discovery of double-stranded RNA-induced gene silencing (RNA interference or RNAi) in *C. elegans* (Fire et al., 1998) precipitated the recognition of the connection between RNAi and miRNAs providing reference to related phenomena in plants and fungi. The short RNAs of a specific length (20–30 nt) and Argonaute proteins manifested the newly found connections (reviewed in Czech & Hannon, 2011; Hutvagner & Simard, 2008) and represented the culmination of the first decade of RNA-silencing research.

From that point, numerous avenues of short RNA studies emerged, and the second decade of RNAi-related research featured advances in biochemistry, structural biology, and genetics. miRNA-based regulation proved to be almost as common as regulation by transcription factors. Advances in deep sequencing technology brought short RNA discovery to new levels and enabled identification of another class of RNAs, piRNAs, which interact with the PIWI subfamily of Argonaute proteins (reviewed in Ishizu, Siomi, & Siomi, 2012). Deep sequencing reinforced genomic approaches in RNAi research in *C. elegans*, and the wealth of endogenous short RNAs (endo-siRNAs), distinct from miRNAs and piRNAs, that came out of these studies is truly remarkable (reviewed in Fischer, 2010). Endo-siRNAs are not unique to *C. elegans*; they exist in *Drosophila* and mammals as well (reviewed in Okamura & Lai 2008), but nematodes developed the most extensive collection of Argonaute proteins, studies of which are poised to discover new biological functions and mechanisms of gene regulation.

This review attempts to cover the breadth of RNAi-related research in *C. elegans* since 2005. For earlier reviews by this author, see Grishok and Mello (2002) and Grishok (2005). miRNAs have received more dedicated reviews than the other short RNAs discovered more recently. Therefore, a more in-depth discussion on miRNA can be found elsewhere (Abbott, 2011; Ambros, 2011; Kaufman & Miska, 2010; Ketting, 2010; Mondol & Pasquinelli, 2012; Pasquinelli, 2012; Ruvkun, 2008; Vella & Slack, 2005). Here, an overview of miRNA research includes their newly discovered biological functions in *C. elegans*, as well as recent advances concerning the molecular mechanisms governing miRNA expression and their role in gene regulation. A large portion of the review is dedicated to the biology of endogenous RNAi in *C. elegans*, including mechanistic insights into the biogenesis of various siRNA species, regulation of mRNA stability and transcription, RNAi inheritance, and connections to other pathways and factors affecting RNAi. The unique biogenesis of *C. elegans* piRNAs (21U-RNAs) is highlighted, as well as their cooperation with endo-siRNAs in genome surveillance. Finally, advances in our understanding of the systemic nature of RNAi in *C. elegans* are described.

2. miRNA FUNCTION

2.1. Biological effects of miRNAs

2.1.1 Developmental timing

Cell divisions and cell-fate decisions during *C. elegans* development follow a stereotypical lineage (reviewed in Ambros, 2011; Rougvie, 2005), which includes programs specific for sequential postembryonic larval stages L1, L2, L3, and L4 and terminal differentiation in the adult. Mutant screens have identified animals with abnormalities in developmental timing, that is, heterochronic mutants, some of which never progress to later developmental programs, while others execute them precociously (reviewed in Ambros, 2011; Rougvie, 2005). The discovery of the first miRNA, *lin-4*, was made during the process of cloning of a gene with a heterochronic mutant phenotype and with the realization that the product of the gene represents a noncoding RNA that exists as a longer hairpin (70 nt) or a shorter (~22 nt) species (Lee et al., 1993). Moreover, genetic analyses revealed that the *lin-14* gene, which itself is important in the heterochronic pathway, is negatively regulated by *lin-4*, and molecular characterization of *lin-14* gain-of-function mutations pointed to the *lin-14* 3′UTR as the site of negative regulation by *lin-4* (Wightman, Ha, & Ruvkun, 1993). After the nature of *lin-4* was revealed, several regions of partial complementarity between *lin-4* RNA and the *lin-14* 3'UTR were identified, and the paradigm of negative regulation by miRNA was born (Lee et al., 1993; Wightman et al., 1993). This seminal work by Victor Ambros, Gary Ruvkun, and their colleagues began to be fully appreciated only by the turn of the twenty-first century when another heterochronic gene, *let-7*, turned out to encode a short RNA (Reinhart et al., 2000), which was remarkably conserved in animals (Pasquinelli et al., 2000), and numerous other endogenous short RNAs with hairpin precursors were cloned from *C. elegans*, *Drosophila*, and mammalian cells (Lagos-Quintana, Rauhut, Lendeckel, & Tuschl, 2001; Lau, Lim, Weinstein, & Bartel, 2001; Lee & Ambros, 2001).

Although mutations in the *lin-4* and *let-7* genes caused obvious developmental phenotypes, extensive analyses of new miRNAs in *C. elegans* concluded that this is not generally true (Miska et al., 2007) and that many miRNAs, especially related ones, work redundantly (Abbott et al., 2005; Alvarez-Saavedra & Horvitz, 2010). For example, *let-7* family members *mir-48*, *mir-84*, and *mir-241* cooperate in repressing the transcription factor Hunchback-like (HBL-1) (Abbott et al., 2005; Li, Jones-Rhoades, Lau, Bartel, & Rougvie, 2005). Nevertheless, control of developmental timing remains the best understood biological function of miRNAs in *C. elegans* (reviewed in Ambros, 2011; Resnick, McCulloch, & Rougvie, 2010; Sokol, 2012), and *let-7* miRNA and its relatives are the most well characterized of all the miRNAs. Their *C. elegans* targets, such as LIN-28, LIN-41, and LET-60/Ras, proved to be conserved in mammals and enabled the recognition of *let-7* as a tumor suppressor and a prodifferentiation factor (reviewed by Ambros, 2011; Mondol & Pasquinelli, 2012).

2.1.2 Embryonic development

The similarity between the RNAi and miRNA pathways in *C. elegans* was recognized because downregulation by RNAi of Argonaute-like genes *alg-1* and *alg-2* resulted in heterochronic phenotypes that resembled the *let-7* mutant (Grishok et al., 2001). Interestingly, the most severe phenotype resulting from inactivation of *alg-1* and *alg-2* is embryonic lethality (Grishok et al., 2001; Vasquez-Rifo et al., 2012), which suggests that some miRNAs or their combination has a role in embryonic development.

Although initial analysis of deletion mutations in 87 miRNA genes identified few mutants with gross developmental abnormalities (Miska et al., 2007), it reported temperature-sensitive embryonic lethality associated with the *mir-35–41* miRNA cluster (Miska et al., 2007). Subsequent deletion of all eight members of the *mir-35* family (*mir-35* through *mir-42*) revealed a more penetrant embryonic phenotype that was not temperature sensitive (Alvarez-Saavedra & Horvitz, 2010). Importantly, this family of miRNAs is expressed during both oogenesis and embryogenesis (Alvarez-Saavedra & Horvitz, 2010; Wu et al., 2010), and the lethality could be rescued by either maternal or early zygotic expression of the miRNAs (Alvarez-Saavedra & Horvitz, 2010). The targets of this miRNA family responsible for the developmental phenotypes could not be identified in suppressor screens and remain unknown.

Another group of related miRNAs required for embryonic development is the *C. elegans mir-51* family (Alvarez-Saavedra & Horvitz, 2010; Shaw, Armisen, Lehrbach, & Miska, 2010), which belongs to a broader miR-100 family defined by human miR-100. The miRNAs of the *mir-51* family, *mir-51–56*, are broadly expressed at all developmental stages (Lim et al., 2003; Shaw et al., 2010; Wu et al., 2010), consistent with their role in embryonic and postembryonic development (Alvarez-Saavedra & Horvitz, 2010; Shaw et al., 2010) and their genetic interaction with multiple developmental pathways (Brenner, Kemp, & Abbott, 2012). The earliest developmental phenotype of *mir-51* family mutants is an unattached pharynx, that is, a lack of attachment of the pharyngeal muscle to the mouth (Shaw et al., 2010). This phenotype was shown to be at least in part due to the misregulation of the *mir-51* target gene, *cdh-3* (Fat cadherin ortholog-3), which is presumed to interfere with the homophilic interactions of another Fat cadherin ortholog required for the maintenance of pharynx attachment, CDH-4 (Schmitz, Wacker, & Hutter, 2008).

As miRNAs of the same family often work redundantly (see earlier), it is easy to imagine that unrelated miRNAs may also cooperate in gene regulation or that miRNAs may act redundantly with other factors. Indeed, Abbot and colleagues found that mutations in 25 out of 31 tested miRNAs showed phenotypes, including embryonic lethality, in sensitized genetic backgrounds (Brenner, Jasiewicz, Fahley, Kemp, & Abbott, 2010).

Notably, in addition to *lin-4* and *let-7*, another miRNA, *lsy-6*, was identified in a forward mutant screen (Johnston & Hobert, 2003). This genetic screen was aimed at finding genes controlling asymmetry of the ASE gustatory neurons, which are morphologically symmetrical but exhibit functional asymmetry (Hobert, 2006). *lsy-6* miRNA is specifically expressed in the left ASE neuron, where it inhibits the transcription factor COG-1 and reinforces expression of asymmetric genes (Johnston & Hobert, 2003). It was discovered recently that *lsy-6* is the first gene to be expressed asymmetrically in the ASE neurons and that this asymmetry is initiated by the TBX-37/38 transcription factors (Cochella & Hobert, 2012). These transcription factors act early in the ASEL but not in the ASER lineage and contribute to chromatin decompaction and "priming" of the *lsy-6* locus for subsequent full activation in ASEL several cell divisions later (Cochella & Hobert, 2012).

2.1.3 Postembryonic development

The miRNAs of the heterochronic pathway, *lin-4* and the *let-7* family, have a profound role in regulating postembryonic development (reviewed in Ambros, 2011; Mondol & Pasquinelli, 2012; Rougvie, 2005; Sokol, 2012). This was first recognized during studies of the defects in lateral hypodermal seam cell lineages (reviewed in Ambros, 2011; Rougvie, 2005). Later, the role of *let-7* family miRNAs in promoting the cessation of molting through inhibition of conserved nuclear hormone receptor genes was described (Hayes, Frand, & Ruvkun, 2006). This regulation takes place through both the heterochronic pathway targets of the *let-7* family, *lin-41* and *hbl-1*, and the direct inhibition of nuclear receptor genes by the *let-7* family miRNAs (Hayes et al., 2006). *lin-4* and the *let-7* family also have important roles during vulva development. The conserved EGF/RAS and Notch signaling pathways govern vulva specification during larval stage 3 (reviewed in Gupta, Hanna-Rose, & Sternberg, 2012). The *let-7* family was first implicated in the direct inhibition of *let-60*/RAS in studies of vulva development in *C. elegans* (Johnson et al., 2005). Moreover, it has been demonstrated recently that the timing of *lin-14* inhibition by *lin-4* in specific vulva precursor cells contributes to the precision of cell-fate specification (Li & Greenwald, 2010).

Also, the timing of nervous system development is controlled by *lin-4* as the negative regulation of *lin-14* and *lin-28* by *lin-4* in the hermaphrodite-specific neuron during larval stage 4 contributes to the timing of its axon elongation (Olsson-Carter & Slack, 2010).

In addition to loss-of-function studies, analyses of miRNA expression patterns using reporters combined with overexpression experiments yielded insight into the role of miRNAs in specific cells. Thus, *mir-61* transcription is directly activated by LIN-12/Notch in specific vulva precursor cells (Yoo & Greenwald, 2005). This miRNA was shown to promote LIN-12 expression in a positive-feedback loop by directly inhibiting its negative regulator *vav-1* (Yoo & Greenwald, 2005). The function of the conserved miRNA *mir-57* in regulation of positional cell-fate specification was identified because of the intriguing posterior expression of an *mir-57* transcriptional reporter throughout embryonic and postembryonic development (Zhao et al., 2010). Although *mir-57* loss-of-function produced weak phenotypes, overexpression of *mir-57* resulted in robust posterior abnormalities (Zhao et al., 2010). Interestingly, *mir-57* was found to directly inhibit the Hox gene *nob-1*, while expression of *mir-57* was dependent on an earlier onset of *nob-1* expression in posterior lineages (Zhao et al., 2010). Another conserved miRNA, *mir-124*, is expressed specifically in ciliated sensory neurons and was shown to promote the mRNA expression signature associated with this type of neuron (Clark et al., 2010).

A number of postembryonic defects, including small body size and defective egg laying, have been reported in mutants lacking all members of the abundantly expressed *mir-58* family (*mir-58*, *mir-80*, *mir-81*, and *mir-82*) (Alvarez-Saavedra & Horvitz, 2010); the targets of this family of miRNAs remain to be found. Also, analyses of miRNA function in sensitized backgrounds have implicated a number of miRNAs in regulation of embryogenesis, adult viability, and the process of gonad migration (Brenner et al., 2010). Despite these efforts, it looks like much of the gene regulation by miRNAs during development remains to be revealed.

2.1.4 Physiology

There has been significant progress in elucidating the physiological roles of miRNAs in *C. elegans*. *mir-1* is an miRNA with conserved expression in muscle tissue. Its elimination in flies and mice is lethal (Sokol & Ambros, 2005; Zhao et al., 2007), while *C. elegans mir-1* mutants are viable. The viability of *mir-1* mutants in *C. elegans* enabled the dissection of its role in synaptic transmission at neuromuscular junctions (Simon et al., 2008). It was shown that *mir-1* modulates muscle sensitivity to acetylcholine through a direct regulation of the abundance of acetylcholine receptor subunits (Simon et al., 2008). Moreover, *mir-1* was implicated in acetylcholine release from neurons due to its nonautonomous role in regulating the transcription factor MEF-2 in the muscle and MEF-2-dependent retrograde signal from the muscle to the neuron (Simon et al., 2008).

The function of another conserved miRNA, *mir-34*, was connected to the regulation of DNA damage–induced cell death in *C. elegans* when *mir-34* mutant animals showed increased susceptibility to radiation in the soma and increased resistance to radiation in the germline (Kato et al., 2009). The mechanistic aspects of *mir-34* function in the DNA damage response are not known.

The role of the *mir-240/786* cluster in the regulation of the *C. elegans* defecation cycle was noted in the initial analysis of a panel of deletion mutants (Miska et al., 2007). A more detailed study of the phenotype revealed that *mir-786*, but not *mir-240*, was involved (Kemp et al., 2012). Moreover, the authors showed that *mir-240/786* is expressed in the posterior intestinal cells, where the pacemaker for the ~50-s calcium-induced defecation rhythm is located, and that *mir-786* directly regulates the fatty acid elongase *elo-2* to control the defecation cycle (Kemp et al., 2012). These results suggest that fatty acid composition in the posterior intestinal cells affects the pacemaker properties of these cells (Kemp et al., 2012).

2.1.5 Longevity and stress response

C. elegans is the first model organism where the genetic pathway of insulin/insulin-like growth factor 1 signaling was shown to negatively regulate life span by inhibiting the longevity-promoting transcription factor DAF-16/FOXO (reviewed in Lapierre & Hansen, 2012). Interestingly, the founding miRNA *lin-4* and its target *lin-14* were also implicated in the DAF-16-dependent control of aging in *C. elegans* either in parallel or through insulin signaling; *lin-4* mutants had a reduced life span, while *lin-14* loss-of-function extended it (Boehm & Slack, 2005). Moreover, *let-7* family members, *mir-84* and *mir-241*, were recently shown to participate in the DAF-16-dependent life span extension induced by germline ablation (Shen, Wollam, Magner, Karalay, & Antebi, 2012). This miRNA-dependent life span increase was due to direct inhibition of *lin-14* and *akt-1* and activation of DAF-16 and its transcriptional targets (Shen et al., 2012). These results are consistent with earlier findings implicating *lin-14* in the negative regulation of longevity (Boehm & Slack, 2005) and also with reports describing a general reduction in miRNA levels, including those of *lin-4* and the *let-7* family, during aging (de Lencastre et al., 2010; Ibanez-Ventoso et al., 2006; Kato, Chen, Inukai, Zhao, & Slack, 2011). In line with this, recent identification of a temperature-dependent mutant allele in the miRNA-processing factor *pash-1* enabled experiments in which miRNA levels were manipulated in adult animals. These experiments revealed that global miRNA reduction accelerates aging (Lehrbach et al., 2012).

Although miRNAs are generally less abundant in aged animals, some miRNAs, such as *mir-34*, *mir-71*, *mir-238*, *mir-239*, and *mir-246*, are upregulated (de Lencastre et al., 2010; Ibanez-Ventoso et al., 2006; Kato et al., 2011). All these miRNAs have been implicated in the control of longevity: *mir-71*, *mir-238*, and *mir-246* act to enhance life span (Boulias &

Horvitz, 2012; de Lencastre et al., 2010), while *mir-34* and *mir-239* inhibit longevity (de Lencastre et al., 2010; Yang et al., 2013). The increased life span of *mir-34* mutants depends on autophagy (Yang et al., 2013) and *mir-239* loss-of-function extends the life span in a DAF-16-dependent manner (de Lencastre et al., 2010). Interestingly, *mir-34* and *mir-71* were also upregulated during starvation-induced L1 diapause and in adults with postdauer life history (Karp, Hammell, Ow, & Ambros, 2011). Increase in life span often correlates with enhanced resistance to stress, while short-lived animals are stress sensitive. Indeed, this correlation was shown for *mir-71*, *mir-238*, *mir-239*, and *mir-246* (Boulias & Horvitz, 2012; de Lencastre et al., 2010).

An in-depth analysis of *mir-71* function determined that *mir-71* is specifically required for life span extension induced by germ cell loss, but not for the increased longevity of animals with reduced insulin signaling (Boulias & Horvitz, 2012). Moreover, although *mir-71* is broadly expressed, it functions in neurons to promote the transcriptional activity of DAF-16 in the intestine of germline-deficient animals (Boulias & Horvitz, 2012).

As studies of miRNAs in aging continue to progress, more similarities between different species emerge. For in-depth reviews on the role of short RNAs in longevity, see Ibanez-Ventoso and Driscoll (2009), Smith-Vikos and Slack (2012), and Kato and Slack (2013), and for more general reviews on the biological functions of miRNAs in *C. elegans*, see Kaufman and Miska (2010) and Abbott (2011).

2.2. Biogenesis and molecular mechanisms of miRNAs

The topic of miRNA biogenesis and their molecular function in *C. elegans* and other species has been reviewed extensively over the years, with a number of excellent reviews published recently (Bartel, 2009; Fabian, Sonenberg, & Filipowicz, 2010; Hammell, 2008; Kai & Pasquinelli, 2010; Kim, Han, & Siomi, 2009; Krol, Loedige, & Filipowicz, 2010; Mondol & Pasquinelli, 2012; Turner & Slack, 2009), including several chapters in the book *Regulation of microRNAs, Advances in Experimental Medicine and Biology* (2010) (Grosshans & Chatterjee, 2010; Ketting, 2010; Lehrbach & Miska, 2010). A summary of the state of the field is provided below without references to primary literature, which can be found in the listed reviews.

miRNAs are generally encoded by RNA polymerase II-transcribed genes, which are subject to positive and negative regulation by specific *cis*-acting transcription factors. Interestingly, miRNAs and the transcription factors that regulate them are often engaged in feedback loops. Primary miRNA transcripts (pri-miRNAs) are capped and polyadenylated; their processing in the nucleus is mediated by the RNase III family enzyme Drosha and its partner Pasha (DRSH-1 and PASH-1, respectively, in *C. elegans*). Some miRNAs are encoded in the introns of protein-coding genes and rely on splicing, not Drosha, for their initial processing in the nucleus. The product of the initial processing, pre-miRNA, is generally a ~70-nt hairpin structure and is further processed in the cytoplasm by another RNase III homolog, Dicer (DCR-1 in *C. elegans*). The product of Dicer processing is a ~22-bp RNA duplex, which consists of a mature miRNA and its complementary "miRNA-star." Generally, the 22-nt strand exhibiting weaker thermodynamics in base pairing at the 5′ end is the mature miRNA. The mature miRNA gets incorporated into the RNA-induced-silencing complex (RISC) as a binding partner of an Argonaute protein. In *C. elegans*, 2 out of 26 Argonaute proteins, ALG-1 and ALG-2, interact with miRNAs (Table 1.1). How these proteins are recruited into the miRNA pathway as opposed to multiple other RNAi-related pathways is not clear, although the structural properties of the hairpin precursor were shown to have a role. Interaction with Argonaute proteins as well as engagement in base pairing with target mRNA protects miRNAs from degradation by exonucleases, such as XRN-2 in *C. elegans*. In recent years, regulation of miRNA stability was found to play an important role in the dynamic properties of miRNA-based gene silencing.

miRNAs regulate their target mRNAs through base pairing at 3′UTRs and by recruiting proteins that ultimately cause translational repression or deadenylation. The key interacting partners of Argonaute proteins that are essential for miRNA silencing are the *C. elegans* homologs of GW182: AIN-1 and AIN-2. miRNAs recognize their targets with imperfect complementarity, and interactions involving the "seed" region, that is nucleotides 2–8, in the mature miRNA occur most often, although other interaction modes have been described for biologically relevant miRNA/mRNA pairs as well. Because of the limited base pairing, computational predictions of valid miRNA targets include other parameters, such as the conservation and accessibility of target sites and the thermodynamics of miRNA–mRNA interaction. In addition to computationally predicted miRNA targets, genomic data on ALG-1 RNA-binding sites and AIN-1/AIN-2 interacting mRNAs are currently available for *C. elegans* researchers, and a targeted proteomic approach has been developed for validation of predicted targets.

Table 1.1 The list of *C. elegans* Argonaute proteins

Protein (Cosmid gene name)	Pathway/function	References
RDE-1 (K08H10.7)	Exo-RNAi, endo-RNAi Slicing of the passenger strand in primary siRNA duplex	Tabara et al. (1999), Steiner, Okihara, Hoogstrate, Sijen, and Ketting (2009), and Correa, Steiner, Berezikov, and Ketting (2010)
ERGO-1 (R09A1.1)	Endo-RNAi (Eri 26G, embryo) Slicing of the passenger strand in primary siRNA duplex	Yigit et al. (2006), Han et al. (2009), and Vasale et al. (2010)
ALG-1 (F48F7.1)	miRNA	Grishok et al. (2001), Hutvagner, Simard, Mello, and Zamore (2004), and Bouasker and Simard (2012)
ALG-2 (T07D3.7)	miRNA	Grishok et al. (2001), Hutvagner et al. (2004), and Bouasker and Simard (2012)
ALG-3 (T22B3.2)	Endo-RNAi (Eri 26G, sperm)	Han et al. (2009) and Conine et al. (2010)
ALG-4 (ZK757.3)	Endo-RNAi (Eri 26G, sperm)	Han et al. (2009) and Conine et al. (2010)
PRG-1 (D2030.6)	21U-RNA (piRNA)	Cox et al. (1998), Batista et al. (2008), Das et al. (2008), and Wang and Reinke (2008)
PRG-2 (C01G5.2)	21U-RNA (piRNA)	Cox et al. (1998), Batista et al. (2008), Das et al. (2008), and Wang and Reinke (2008)
CSR-1 (F20D12.1)	Slicer with secondary siRNAs Endo-RNAi (22G, antisense to genes) Germline and soma, nuclear and cytoplasmic	Yigit et al. (2006), Aoki, Moriguchi, Yoshioka, Okawa, and Tabara (2007), and Claycomb et al. (2009)
C04F12.1	Most closely related to CSR-1	Yigit et al. (2006) and Gu et al. (2009)

Continued

Table 1.1 The list of *C. elegans* Argonaute proteins—cont'd

Protein (Cosmid gene name)	Pathway/function	References
NRDE-3/ WAGO-12 (R04A9.2)	Nuclear exo-RNAi, secondary siRNAs Soma endo-RNAi (22G, ERGO-1 26G-dependent)	Yigit et al. (2006), Guang et al. (2008), and Gu et al. (2009)
HRDE-1/ WAGO-9 (C16C10.3)	Heritable exo-RNAi Nuclear endo-RNAi (22G, germline) 21U-initiated 22G	Yigit et al. (2006), Gu et al. (2009), Ashe et al. (2012), Buckley et al. (2012), and Shirayama et al. (2012)
WAGO-1 (R06C7.1)	Exo-RNAi, secondary siRNA Cytoplasmic endo-RNAi (22G, Eri, repeats), 21U-initiated 22G	Yigit et al. (2006), Gu et al. (2009), and Shirayama et al. (2012)
WAGO-2 (F55A12.1)	Exo-RNAi, secondary siRNA Endo-RNAi (22G, Eri, repeats)	Yigit et al. (2006) and Gu et al. (2009)
WAGO-3/ PPW-2 (Y110A7A.18)	Exo-RNAi, secondary siRNA Endo-RNAi (22G, Eri, transposons, repeats)	Vastenhouw et al. (2003), Yigit et al. (2006), and Gu et al. (2009)
WAGO-4 (F58G1.1)	Exo-RNAi, secondary siRNA Endo-RNAi (22G, Eri, repeats)	Yigit et al. (2006) and Gu et al. (2009)
WAGO-5 (ZK1248.7)	Exo-RNAi, secondary siRNA Endo-RNAi (22G, Eri, repeats)	Yigit et al. (2006) and Gu et al. (2009)
WAGO-6/ SAGO-2 (F56A6.1)	Exo-RNAi, secondary siRNA Endo-RNAi (22G, Eri, repeats)	Yigit et al. (2006) and Gu et al. (2009)
WAGO-7/ PPW-1 (C18E3.7)	Exo-RNAi, secondary siRNA Endo-RNAi (22G, Eri, repeats)	Tijsterman, Okihara, Thijssen, and Plasterk (2002), Yigit et al. (2006), and Gu et al. (2009)
WAGO-8/ SAGO-1 (K12B6.1)	Exo-RNAi, secondary siRNA Endo-RNAi (22G, Eri, repeats)	Yigit et al. (2006) and Gu et al. (2009)
WAGO-10 (T22H9.3)	Exo-RNAi, secondary siRNA Nuclear endo-RNAi (22G, Eri, repeats) 21U-initiated 22G	Yigit et al. (2006), Gu et al. (2009), and Shirayama et al. (2012)

Table 1.1 The list of *C. elegans* Argonaute proteins—cont'd

Protein (Cosmid gene name)	Pathway/function	References
WAGO-11 (Y49F6A.1)	Exo-RNAi, secondary siRNA Endo-RNAi (22G, Eri, repeats)	Yigit et al. (2006) and Gu et al. (2009)
T23D8.7	ALG-brunch of Argonautes	Yigit et al. (2006)
C14B1.7	WAGO-brunch of Argonautes	Yigit et al. (2006)
C06A1.4	Pseudogene?	Yigit et al. (2006)
M03D4.6	Pseudogene?	Yigit et al. (2006)

Identification of biologically relevant miRNA targets and pathways regulated by miRNAs is challenging, but it continues to grow quickly. Also, in addition to the main components involved in miRNA biogenesis and function, highlighted earlier, a number of their cofactors and modulators have been identified. The emerging picture is that silencing by miRNAs is affected by cell-specific factors that either enhance or relieve the effect of miRNAs by a variety of mechanisms. These include regulation of transcription, processing or stability of miRNAs, a cooperation between miRNAs and other RNA-binding proteins in the regulation of mRNA stability or translation, as well as regulation of alternative polyadenylation and, consequently, 3′UTR length, which often eliminates miRNA binding and silencing. Surprisingly, the nuclear function of *let-7* miRNA and ALG-1 in promoting the processing of *let-7* pri-mRNA has been described recently (Zisoulis, Kai, Chang, & Pasquinelli, 2012), and a role for zinc finger protein SOMI-1 and chromatin-remodeling factors in promoting *mir-84*-dependent gene regulation was reported (Hayes, Riedel, & Ruvkun, 2011). These results suggest that miRNAs may exist in complexes distinct from RISC and guide the activity of nuclear RNA-processing and, perhaps, that of chromatin-binding factors.

3. RNAi AND ENDOGENOUS siRNAs

Initial studies of RNAi in *C. elegans* that were conducted before 2005 (see earlier reviews by this author (Grishok, 2005; Grishok & Mello, 2002), and references therein) defined its key properties, such as dependence on dsRNA, systemic nature and heritability, and also discovered genes required for RNAi. These studies also recognized that some of the genes, such as *rde-1* and *rde-4*, act to initiate the silencing response while others, such as *mut-7* and *rde-2*, work at more downstream steps, and that yet another group of factors, such as *sid-1*, is involved in the systemic transport of silencing. Also, at the turn of the twenty-first century, siRNAs were detected in *C. elegans*, Dicer-dependent processing of dsRNA to siRNAs was confirmed using *C. elegans* extracts, and amplification of silencing RNA agents was suggested by the discovery of the role of RNA-dependent RNA polymerases (RdRP) in RNAi. Moreover, during these years, the similarity between RNAi and other homology-dependent silencing phenomena was recognized, as well as the antagonistic relationship between RNAi and adenosine deaminases that act on RNA (ADARs). Finally, the identification of mutants, such as those inactivating RdRP *rrf-3*, which were more susceptible to exogenous RNAi, together with the discovery of endogenous short RNAs that were distinct from miRNAs, predicted the existence of endogenous RNAi-based mechanisms. Similar to the discovery of miRNAs, endo-siRNAs were initially found in *C. elegans* and later identified in flies and mammals.

3.1. Exogenous RNAi pathway

3.1.1 RDE-4 and Dicer

The upstream role of the *rde-4* gene in the RNAi pathway was validated by the molecular properties of its protein product. *rde-4* encodes a protein with two dsRNA-binding motifs (dsRBM1 and dsRBM2) and was shown to bind long dsRNA and to exist in a complex with Dicer (DCR-1), dicer-related helicase DRH-1 and Argonaute protein RDE-1 (Tabara, Yigit, Siomi, & Mello, 2002; Figure 1.1). This complex is involved in the initiation of the silencing response to exogenous dsRNA. Its existence was further validated by proteomic analyses of the interacting partners of DCR-1 (Duchaine et al., 2006) and by characterization of DCR-1-containing complexes using gel filtration (Thivierge et al., 2012). RDE-4 was shown to bind long dsRNA with much higher affinity than short (~20 bp) dsRNA species and to dimerize through its C-terminal domain (Parker, Eckert, & Bass, 2006). Interestingly, although dimerization of RDE-4 was not required for dsRNA binding, it was important for dsRNA processing, as shown by *in vitro* siRNA production assays with *rde-4* mutant extracts supplemented with recombinant RDE-4 (Parker et al., 2006). A subsequent *in vitro* study demonstrated that RDE-4 binds dsRNA cooperatively, and that dsRBM2 is important for dsRNA binding (Parker, Maity, & Bass, 2008). It also showed that the linker region between the two dsRBMs is required for dsRNA

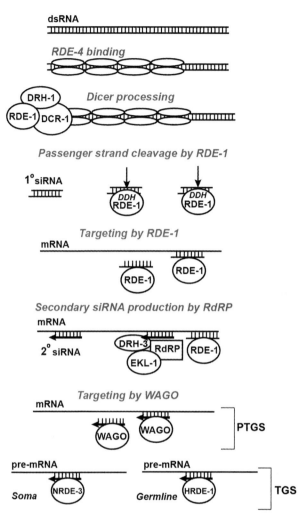

Figure 1.1 Schematic of the exogenous RNAi pathway, which responds to environmental or injected dsRNA. The RDE-4 dimer binds dsRNA followed by the Dicer complex containing Dicer-related helicase DRH-1 and Argonaute RDE-1. Dicer cleaves dsRNA to produce primary double-stranded siRNAs bound by RDE-1 (shown with catalytic amino acid triad DDH). RDE-1 cleaves the passenger strand in the siRNA duplex and is brought to the target mRNA by the guide strand. Targeting by RDE-1 stimulates secondary siRNA production by the RdRP (RRF-1 or EGO-1) complex using selected mRNA as a template. WAGO Argonautes bind to the secondary siRNAs and initiate posttranscriptional gene silencing (PTGS) through poorly understood mechanisms. Somatic NRDE-3/WAGO-12 binds to secondary siRNAs in the cytoplasm and brings them to the nucleus where they target pre-mRNA and induce transcriptional gene silencing (TGS). Nuclear Argonaute HRDE-1/WAGO-9 mediates TGS in the germline. Only factors with clear mechanistic roles are included in the schematic (here and in other figures).

processing and predicted that this region interacts with Dicer (Parker et al., 2008). *In vivo* studies confirmed the important functional roles of dsRMB2 and the linker region as well as their requirement for the interaction between RDE-4 and DCR-1 (Blanchard et al., 2011). It is thought that RDE-4 recognizes dsRNA and recruits Dicer and the rest of the complex for dsRNA processing into siRNAs (Parker et al., 2006; Figure 1.1). Interestingly, it was reported that when a high level of dsRNA is produced from transgenes, initiation of the silencing process does not require RDE-4 (Habig, Aruscavage, & Bass, 2008). It is not clear whether the level or the timing of RDE-4 expression is extensively regulated as it mediates RNAi at all developmental stages and in both somatic and germ cells. In the germline, a new protein, DEPS-1, which is required for the assembly of P-granules in germ cells, was found to promote *rde-4* mRNA and protein expression in this tissue and, consequently, efficient RNAi response in the germline (Spike, Bader, Reinke, & Strome, 2008). The role of DEPS-1 in control of *rde-4* is separate from its function in P-granule assembly. In general, RDE-4 acts nonautonomously in RNAi (Blanchard et al., 2011; Jose, Garcia, & Hunter, 2011). Therefore, its regulation in one tissue may affect the efficiency of RNAi in another and its regulation in the germline may affect the progeny.

While most of the biochemical and structural work on Dicer has been conducted in other systems, an important insight into the role of its helicase domain was achieved using the *dcr-1(mg375)* mutant with a point mutation in this domain (Pavelec, Lachowiec, Duchaine, Smith, & Kennedy, 2009; Welker et al., 2011, 2010). Although DCR-1 helicase mutations primarily affected endogenous siRNA levels (Pavelec et al., 2009; Welker et al., 2010) and did not seem to affect exogenous RNAi (Welker et al., 2010) in these *in vivo* studies, it was discovered later that the helicase domain is important for the generation of siRNAs from internal regions of long dsRNA with blunt termini because *dcr-1(mg375)* mutant extracts were deficient in this process (Welker et al., 2011). miRNA precursors do not require such processive action of Dicer, and miRNA levels were not affected by mutations in its helicase domain (Welker et al., 2010, 2011).

3.1.2 RDE-1 Argonaute protein

Argonaute proteins exist in complexes with short RNAs and function in RNA-mediated silencing processes (reviewed in Czech & Hannon, 2011; Hutvagner & Simard, 2008). There are 26 genes encoding Argonaute family proteins in *C. elegans* (Table 1.1), but only one, *rde-1*, was implicated in the exogenous RNAi pathway through unbiased discovery in the initial screen for RNAi-deficient mutants (Tabara et al., 1999). Although RDE-1 is not required for dsRNA processing *per se* (Parrish & Fire, 2001), it interacts with the Dicer complex (Duchaine et al., 2006; Tabara et al., 2002) and binds to the primary siRNA duplex generated by Dicer (Yigit et al., 2006; Figure 1.1).

miRNAs are naturally processed from short hairpins with ~22 bp stems, where bulges disrupting pairing are common, and siRNAs are generated from long dsRNA precursors in exogenous RNAi. However, it has been shown that hairpin precursors with perfectly base-paired stems are efficiently processed into siRNAs in mammalian cells (reviewed in McManus, Petersen, Haines, Chen, & Sharp, 2002; McManus & Sharp, 2002) and in *C. elegans* (Sijen, Steiner, Thijssen, & Plasterk, 2007). There are no dedicated Argonaute proteins for RNAi and miRNAs in mammals, but there is a clear separation between Argonautes acting in the exo-RNAi and miRNA pathways in *C. elegans*: RDE-1 is specific for RNAi (Tabara et al., 1999), while ALG-1 and ALG-2 are specific for miRNAs (Grishok et al., 2001; Table 1.1). Interestingly, it was shown that perfect base pairing in siRNA hairpins was important for the loading of the processed product onto RDE-1, while precursors containing bulges were loaded onto ALG-1 (Jannot, Boisvert, Banville, & Simard, 2008; Steiner et al., 2007). However, this distinction is not perfect, as many natural miRNA precursors contain few mismatches and their resulting mature miRNAs have been found in complex with RDE-1 (Steiner et al., 2007). In fact, miRNAs comprise the majority of short RNAs bound to RDE-1 in wild-type worms that are not subject to exogenous RNAi (Correa et al., 2010). Despite the fact that RDE-1 preferentially binds to short RNAs produced from perfectly base-paired substrates, *let-7* miRNA generated in such a manner requires ALG-1 for its biological function (Jannot et al., 2008). Also, siRNAs targeting the gene *unc-22*, which were generated from a mismatched precursor suited for ALG-1 loading, induced the *unc-22*-silencing phenotype and required RDE-1 for their activity (Steiner et al., 2007). These results indicate that although the structure of the precursor predisposes short RNAs for loading onto Argonautes with appropriate functional roles, it does not determine the functional specificity of the pathway, which is determined by the Argonaute itself, possibly through its interacting partners. Indeed, it was shown that RDE-1 and ALG-1 reside in separate nucleoprotein complexes (Gu et al., 2007; Steiner et al., 2007). Moreover, although artificial siRNAs were produced from the hairpin precursor in *rde-1* and *rde-4* mutant backgrounds, they remained in complex with Dicer (Sijen et al., 2007), suggesting that RDE-4 is required for siRNA loading even though it binds poorly to short dsRNAs (Parker et al., 2006).

Although all Argonaute proteins are capable of binding short RNAs, they differ in their RNase H–like capacity to cleave RNA, which is dependent on the DDH catalytic motif (reviewed in Czech & Hannon, 2011; Hutvagner & Simard, 2008). The catalytic activity of Argonautes from other systems is used in two steps in the RNAi process: (1) removal of the "passenger" strand in the siRNA duplex and (2) degradation of the mRNA target (reviewed in Czech & Hannon, 2011; Hutvagner & Simard, 2008). As RDE-1 was shown to act upstream in the RNAi process, it seemed unlikely that its "slicer" capacity would be used for mRNA degradation. Indeed, in experiments where RDE-1 proteins with mutations in the DDH motif were introduced in an *rde-1* mutant background, it was shown that this motif was required for the RNAi response specifically at the step of passenger-strand cleavage, and that mutant RDE-1 proteins were not capable of binding to the target mRNA (Steiner et al., 2009; Figure 1.1). As RDE-1 is absolutely required for exogenous RNAi and binds primary siRNAs generated by Dicer, it seems to guide the downstream components of the RNAi pathway, such as the RdRP complex, to the target mRNA without causing mRNA cleavage (Figure 1.1).

3.1.3 siRNA amplification and additional Argonautes

The dsRNA-induced silencing effect in *C. elegans* is remarkably strong, although the amount of injected dsRNA is limited (Fire et al., 1998). The possibility of RNAi amplification had been conjectured about early on in the RNAi field and was confirmed with the discoveries of the requirement of the RdRP genes, *ego-1* and *rrf-1*, for exo-RNAi (Sijen et al., 2001; Smardon et al., 2000). Early models envisioned the generation of long dsRNA by RdRPs using target mRNAs as templates and primary siRNAs as primers and the processing of these secondary dsRNA molecules by Dicer (Sijen et al., 2001). However, the discovery that secondary siRNAs are strictly target dependent but largely antisense to mRNA by Northern blotting (A. Grishok, P.D. Zamore and C.C. Mello, unpublished, Figure 1.2) suggested a different mechanism. Indeed, subsequent studies determined that secondary siRNAs are products of *de novo* synthesis by RdRPs (Figure 1.1) and carry 5′triphosphates (Pak & Fire, 2007; Sijen et al., 2007) and identified endogenous siRNA molecules with similar characteristics (Ambros, Lee, Lavanway, Williams, & Jewell, 2003; Lee, Hammell, & Ambros, 2006;

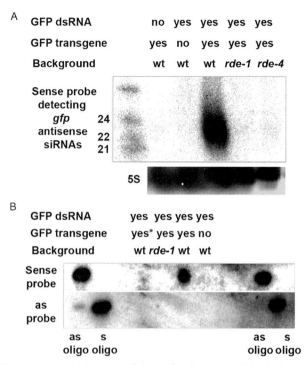

Figure 1.2 Target-dependent accumulation of antisense siRNAs during exo-RNAi. (A) Top: Northern blot detecting antisense *gfp* siRNAs in worms either containing or lacking the *pes-10::gfp* transgene and either exposed to *gfp* dsRNA or not. Only those worms that contain the transgene (target *gfp* mRNA) and that are exposed to *gfp* dsRNA show siRNA accumulation. *rde-1* and *rde-4* are required for the initiation of RNAi response and for secondary siRNA accumulation *in vivo*. Bottom: 5S RNA detection by northern blot is shown as a loading control. (B) Secondary siRNAs are antisense to the target (*gfp*) mRNA and are not detected by the probe designed to detect sense siRNAs. *gfp* sense and antisense RNA oligos are loaded for control of hybridization. Asterisk designates a transgenic strain *gsk-3*::GFP that shows a lower accumulation of siRNAs after dsRNA treatment compared to the *pes-10::gfp* strain (all other GFP transgene lanes in both panels). *Data presented in this figure are from the Ph.D. thesis of Alla Grishok, UMass Worcester, 2001.*

Pak & Fire, 2007; Ruby et al., 2006). Moreover, the production of secondary siRNAs by RdRP activity was recapitulated *in vitro* (Aoki et al., 2007). Interestingly, although secondary siRNAs are very efficient at promoting target mRNA degradation both directly (Aoki et al., 2007) and indirectly (Yigit et al., 2006), their competence in inducing tertiary siRNA production is limited (Montgomery et al., 2012; Pak, Maniar, Mello, & Fire, 2012). Therefore, the primary siRNAs are potent in promoting secondary siRNA production but do not induce silencing themselves, while secondary siRNAs are efficient in executing silencing effects but have a limited capacity to induce indefinite siRNA amplification (Pak et al., 2012).

Although generation of secondary siRNAs was shown to be Dicer independent (Aoki et al., 2007), RRF-3 RdRP was discovered in Dicer immunoprecipitates (Duchaine et al., 2006). It turned out that some of the endogenous RNAi pathways require Dicer processing of dsRNA generated by RRF-3 (see discussion in Section 3.2). A recent biochemical analysis revealed that RdRPs RRF-1 and EGO-1, which are responsible for Dicer-independent production of siRNAs, exist in a complex with Dicer-related helicase DRH-3 (Aoki et al., 2007; Gu et al., 2009; Thivierge et al., 2012) and tudor domain protein EKL-1 (Gu et al., 2009; Thivierge et al., 2012; Figure 1.1), while the RRF-3 complex contains DRH-3, Dicer, and a distinct tudor domain protein, ERI-5 (Thivierge et al., 2012).

The Argonaute protein RDE-1 was implicated in the initiation of the RNAi response in *C. elegans* (Grishok, Tabara, & Mello, 2000; Tabara et al., 1999, 2002), while *Drosophila* Ago2 was identified in an RISC that mediates target mRNA degradation (Hammond, Boettcher, Caudy, Kobayashi, & Hannon, 2001). As the family of Argonaute genes is extensive in *C. elegans* (it includes 26 members, Table 1.1), this suggested that additional Argonaute proteins act downstream in the *C. elegans* RNAi process, similar to *Drosophila* Ago2. Indeed, multiple Argonaute proteins were found to contribute incrementally to the efficiency of RNAi, such that multiple Argonaute (MAGO) mutant strains, which contain up to 12 individual mutant Argonaute genes (e.g., MAGO12), are resistant to RNAi (Gu et al., 2009; Yigit et al., 2006). MAGO Argonautes (also called worm-specific Argonautes or WAGO) are found in complexes with secondary siRNAs produced by RdRPs (Gu et al., 2009; Yigit et al., 2006; Figure 1.1). These downstream Argonautes appear to be limiting because their overexpression leads to a higher sensitivity to RNAi (Yigit et al., 2006). Unexpectedly, WAGO family Argonautes do not contain the amino acid residues required for endonucleolytic activity; therefore, they must recruit additional cofactors for target mRNA degradation (Yigit et al., 2006).

Another Argonaute protein, CSR-1, was shown to promote the efficiency of exogenous RNAi in the germline (Yigit et al., 2006). However, it functions mainly in one of the endogenous RNAi pathways described later in this chapter, and its role in exo-RNAi may be indirect (Gu et al., 2009). Despite the fact that "slicing" by Argonaute proteins does not appear to play a major role in silencing induced by exogenous RNAi, *C. elegans* extracts are fully capable of supporting slicing reactions with exogenously provided siRNAs (Aoki et al., 2007). siRNAs that mimic secondary molecules produced by RdRP were shown to be more efficient in *in vitro* slicing compared to those with primary siRNA features, and the CSR-1 Argonaute was found to be responsible for this Slicer activity (Aoki et al., 2007).

3.1.4 Mutator proteins

The *mut-7* and *rde-2/mut-8* genes were found to act downstream of primary siRNAs in the RNAi pathway (Grishok et al., 2000; Ketting, Haverkamp, van Luenen, & Plasterk, 1999; Ketting & Plasterk, 2000; Tijsterman, Ketting, Okihara, Sijen, & Plasterk, 2002). They belong to a larger group of "mutator" genes that are required for transposon silencing in the germline and for RNAi; mutations in this group of genes cause transposon mobilization and a high frequency of spontaneous mutations (Collins, Forbes, & Anderson, 1989). MUT-7 contains a $3'$–$5'$ exonuclease domain and may participate in target mRNA degradation directly (Ketting et al., 1999). Biochemical analysis of the MUT-7 protein demonstrated that it is present in separate complexes in the nucleus and in the cytoplasm (Tops et al., 2005). The cytoplasmic complex also contains RDE-2 and was shown to increase in size upon dsRNA treatment (Tops et al., 2005). This increase in size was not observed in *rde-1* and *rde-4* mutants (Tops et al., 2005). Although the cytoplasmic MUT-7 complex appears to be distinct from the complex containing RdRP RRF-1 (Thivierge et al., 2012), it is interesting that specific cellular compartments called "*Mutator* foci," identified at the periphery of germline nuclei and containing six mutator proteins, also showed colocalization with RRF-1 (Phillips, Montgomery, Breen, & Ruvkun, 2012). The formation of *Mutator* foci was shown to be dependent on the glutamine/asparagine (Q/N)-rich protein MUT-16; *Mutator* foci also include MUT-7, RDE-2/MUT-8, nucleotidyl transferase MUT-2/RDE-3, DEAD-box RNA helicase MUT-14, and MUT-15/RDE-5 (Phillips et al., 2012). The *Mutator* foci were found to be adjacent to germline P-granules but distinct from them and were found not to depend on P-granule components for their stability (Phillips et al., 2012).

3.1.5 ABC transporters

dsRNA can be introduced into *C. elegans* by injection or by feeding dsRNA-expressing bacteria (environmental RNAi), and the RNAi-silencing process spreads systemically (see discussion in Section 5). Mutants deficient in gene silencing initiated by dsRNA feeding but supporting silencing induced by dsRNA injection were thought to influence the uptake of dsRNA from the environment (Tabara et al., 1999; Tijsterman, May, Simmer, Okihara, & Plasterk, 2004). Surprisingly, the RNAi sensitivity of many of these mutants was shown to depend on the concentration of the dsRNA trigger rather than the method of its delivery, such that the defect in RNAi was revealed only at lower concentrations (Han, Sundaram, Kenjale, Grantham, & Timmons, 2008; Sundaram, Echalier, Han, Hull, & Timmons, 2006; Sundaram et al., 2008). One class of genes implicated in the efficiency of RNAi is the membrane-localized ATP-binding cassette (ABC) transporters, which use ATP to translocate small molecules across membranes (Sundaram et al., 2006, 2008). A mutation in the ABC transporter gene *haf-6* was isolated in a genetic screen (which also yielded *rde-1* and *rde-4*; Tabara et al., 1999) and was characterized later (Sundaram et al., 2006). Following this, 49 out of 61 ABC transporter genes were analyzed for their role in RNAi, and 10 of them, termed ABC$_{RNAi}$, were confirmed to promote the silencing process (Sundaram et al., 2006, 2008). The ABC domain of HAF-6 was shown to be important for its role in RNAi and for the localization of HAF-6 to membrane structures consistent with endoplasmic reticulum in the intestine and germline, two tissues where RNAi defects were observed in *haf-6* mutants (Sundaram et al., 2006).

Interestingly, mutants in ABC$_{RNAi}$ transporters showed genetic interaction with the mutator class of RNAi-resistant mutants: double heterozygotes containing an ABC$_{RNAi}$ mutation and a Mut mutation showed a second-site noncomplementation interaction that resulted in RNAi resistance (Sundaram et al., 2008). Similar genetic interaction was reported previously for *mut-7* and *rde-2/mut-8*, which encode components of the same complex (Tops et al., 2005). Consistently, ABC$_{RNAi}$ mutants show defects in transposon silencing, that is, the mutator phenotype (Sundaram et al., 2008). The genes *rsd-2* and *rsd-6* were also initially thought to be involved in systemic RNAi silencing (Tijsterman et al., 2004), but were later described as dose-dependent mutants with mutator phenotypes (Han et al., 2008). RSD-2 localizes to multiple cellular compartments, including the nucleolus, and partially colocalizes with HAF-6 (Han et al., 2008). *haf-6*, *rsd-2*, and *rsd-6* mutants were shown to be required for the accumulation of some endogenous siRNAs and secondary siRNAs in the exo-RNAi pathway (Zhang et al., 2012). Intriguingly, HAF-6 localization was observed in the perinuclear region in germ cells (Sundaram et al., 2008); it would be interesting to explore whether this localization overlaps with the recently found *Mutator* foci (Phillips et al., 2012).

3.1.6 RDE-10/RDE-11 complex

The RDE-10/RDE-11 complex is the most recently identified protein complex promoting exogenous RNAi (Yang et al., 2012; Zhang et al., 2012). This complex appears to be nematode-specific. The *rde-10* and *rde-11* genes were identified in genetic screens for RNAi-deficient mutants, and their protein products were found to be major interacting partners through proteomic analyses (Yang et al., 2012; Zhang et al., 2012). The RDE-10/RDE-11 complex has been shown to promote secondary siRNA amplification and mRNA degradation and to work in parallel with the nuclear RNAi pathway (Yang et al., 2012; Zhang et al., 2012). In addition, this complex appears to show preference for partially degraded mRNA (Yang et al., 2012). The RDE-10 protein has been found to associate with mRNA targeted by exo-RNAi, and this interaction was shown to be independent of *rrf-1* and *rde-11* but dependent on *rde-1* (Yang et al., 2012). Altogether, these results suggest that the RDE-10/RDE-11 complex engages target mRNA in response to RDE-1 and primary siRNAs but prior to RRF-1 and stabilizes partially degraded mRNA to promote the generation of secondary siRNAs (Yang et al., 2012; Zhang et al., 2012).

3.1.7 Transcriptional silencing induced by dsRNA

The mechanism of gene silencing induced by dsRNA in *C. elegans* is still poorly understood. It appears that posttranscriptional, cotranscriptional, and transcriptional mechanisms are involved and that some genes are preferentially silenced through mRNA degradation in the cytoplasm, while pre-mRNA degradation coupled with transcriptional silencing is prevalent for others. For example, RNAi-induced transcriptional silencing of the *elt-2:: GFP/LacZ* reporter includes a decrease in pre-mRNA levels, reduction in RNA Pol II occupancy at the transgenic array, and a decrease in histone H4 acetylation (Grishok, Sinskey, & Sharp, 2005). In this system, the initiation of silencing was shown to be dependent on *rde-1* and *rde-4*, and downregulation of a number of genes with predicted nuclear roles had an effect on the efficiency of silencing (Grishok et al., 2005).

Detailed analyses of dsRNA-induced silencing of nuclear transcripts such as *lir-1-lin-26* (Bosher, Dufourcq, Sookhareea, & Labouesse, 1999) and *lin-15b-lin-15a* polycistronic RNAs have been published by Kennedy and colleagues (Burkhart et al., 2011; Guang et al., 2010, 2008). Several mutants that specifically affect nuclear RNAi (*nuclear RNAi defective* (*nrde*)) were identified in forward genetic screens, and the molecular functions of the corresponding gene products were investigated. Interestingly, an Argonaute protein, NRDE-3, was shown to be specifically required for nuclear RNAi (Guang et al., 2008; Figure 1.1). NRDE-3 is expressed in somatic cells, contains a nuclear localization signal (NLS), and localizes to the nucleus in an NLS-dependent and siRNA-dependent manner (Guang et al., 2008). NRDE-3 is also called WAGO-12 and belongs to the group of WAGO Argonautes (Gu et al., 2009; Table 1.1), consistent with its binding of secondary siRNAs produced by RdRP RRF-1 (Burkhart et al., 2011; Guang et al., 2008; Figure 1.1). NRDE-3 escorts siRNAs to the nucleus, where it binds target pre-mRNA and promotes the interaction between the nuclear serine/arginine-rich conserved protein NRDE-2 and pre-mRNA. The action of NRDE-3, NRDE-2, and the two additional nematode-specific proteins NRDE-1 and NRDE-4 leads to inhibition of pre-mRNA synthesis and depletion of the RNA Pol II ChIP signal downstream of the region targeted by dsRNA, while Pol II appears to accumulate at DNA sequences corresponding to the dsRNA (Burkhart et al., 2011; Guang et al., 2010). This effect of RNAi on transcriptional elongation is accompanied by an increase in histone H3 lysine 9 (H3K9) methylation near the dsRNA-targeted region (Burkhart et al., 2011; Guang et al., 2010). Notably, the target genes chosen for the exo-RNAi-induced silencing experiments that revealed the function of the NRDE system normally have very low levels of H3K9 methylation and therefore do not represent the endogenous targets of this pathway. Although NRDE-3 and NRDE-2 were shown to interact with pre-mRNA in an siRNA-dependent manner, these proteins do not associate with chromatin, while NRDE-1 does (Burkhart et al., 2011). Interestingly, mutation in *nrde-4* does not prevent the interaction of NRDE proteins with pre-mRNA, but it abolishes NRDE-1 recruitment to chromatin and H3K9 methylation (Burkhart et al., 2011). The NRDE system is likely to provide further insight into the mechanisms of dsRNA-induced H3K9 methylation and transcription inhibition.

3.1.8 Inheritance of dsRNA-induced gene silencing

dsRNA-induced gene silencing was originally shown to occur in animals injected with dsRNA and in their progeny, but it did not persist further (Fire et al., 1998). Moreover, mated *rde-1* and *rde-4* mutants injected with dsRNA were fully capable of transmitting the silencing signal to their *rde/+* progeny, which indicates that generation of primary siRNAs and target-dependent siRNA amplification in P0 is not required for the persistence of the silencing signal in the F1 generation (Tabara et al., 1999). These data are consistent with a persistence of the dsRNA that precedes primary siRNA production (see discussion in Section 5). The dsRNA is likely to be distributed to all cells in the F1 progeny as the zygote divides and to eventually lead to target mRNA-dependent siRNA amplification and silencing in cells expressing the target mRNA. A recent study found that the persistence of dsRNA-induced silencing for target genes expressed in the F1 larva, but not those expressed in the embryo, required the *nrde* pathway genes and that these genes promoted accumulation of NRDE-3-bound siRNAs (Burton, Burkhart, & Kennedy, 2011). Moreover, this study found that F1 animals exhibited a much higher level of NRDE-dependent H3K9 methylation at the dsRNA-targeted locus compared to their parents, which had been exposed to the dsRNA (Burton et al., 2011). These results indicate that the nuclear RNAi pathway and, specifically, the NRDE-3 Argonaute, which associates with secondary siRNAs, are predominantly responsible for silencing in somatic tissues of F1 animals.

Inheritance of RNAi beyond the F1 generation is qualitatively different from the persistence of RNAi in F1 because its initiation requires the function of *rde-1* and *rde-4*, and therefore primary siRNA production, siRNA amplification, or both, in the P0 animals injected with dsRNA (Grishok et al., 2000). The fact that this type of RNAi inheritance is mediated by an extragenic epigenetic factor supports the idea that siRNAs are inherited (Grishok et al., 2000). Moreover, as inheritance of dsRNA-induced silencing beyond the F1 generation has been demonstrated only for genes expressed in the germline (Alcazar, Lin, & Fire, 2008; Buckley et al., 2012; Grishok et al., 2000; Gu et al., 2012; Vastenhouw et al., 2006), secondary siRNAs generated on maternal transcripts and deposited into the zygote are the most likely agents of long-term RNAi inheritance.

Inheritance of RNAi beyond the F1 generation was first shown in experiments targeting essential genes and therefore selecting progeny inheriting less silencing material (Grishok et al., 2000). In a subsequent study, a detailed pedigree-based analysis of dsRNA-induced viability over generations was performed (Alcazar et al., 2008). This study identified a bottleneck in RNAi inheritance at the F4 generation, such that F4 animals survived well but rarely had viable progeny themselves (Alcazar et al., 2008). The properties of RNAi inheritance up to the F4 generation were consistent with that of a diffusible epigenetic element that could be transmitted through both oocyte and sperm (Alcazar et al., 2008).

Genomic analysis was further used for studies of the connection between RNAi and chromatin and transgenerational RNAi inheritance (Gu, Pak, et al., 2012). This work clearly identified a dsRNA-induced H3K9 methylation mark at the several loci targeted by dsRNA in animals fed with dsRNA as well as in their F1 and F2 progeny. Interestingly, H3K9me3 could spread from the targeted locus and was detected up to 11 kb away from the dsRNA trigger region (Gu, Pak, et al., 2012). Notably, dsRNA-induced H3K9me3 required *rde-1*, WAGO secondary Argonautes, and *nrde-2*, even though the level of siRNA production was not altered in *nrde-2* mutant worms (Gu, Pak, et al., 2012). siRNAs were found to decrease in abundance in the F1 and F2 generations, although they were still present; therefore, this study is consistent with the idea of inherited siRNAs and suggests that they induce H3K9me3 (Gu, Pak, et al., 2012). Importantly, although the accumulation of siRNAs was detected 4 h after dsRNA exposure and increased further after 24 h, no H3K9 methylation was present at the target loci at that time, appearing only after 48 hours (Gu, Pak, et al., 2012). This significant lag period between the two events argues against immediate siRNA-guided modification of chromatin and suggests multiple intermediate steps in this process. Another study of dsRNA-induced heritable silencing of a single copy GFP reporter followed the persistence of silencing in more than 60% of the animals over four generations (Ashe et al., 2012). Although both sense and antisense siRNAs were identified in animals treated with dsRNA, only antisense siRNAs persisted to the F4 generation (Ashe et al., 2012). Consistent with the detection of H3K9 methylation at the loci targeted by RNAi described earlier, a mutation in the gene *set-25*, which encodes a putative H3K9 methyltransferase, alleviated the heritable silencing induced by dsRNA (Ashe et al., 2012).

Similar to the requirement of the *nrde* pathway genes *nrde-1*, *nrde-2*, and *nrde-4* for the persistence of dsRNA-induced silencing that targets somatic genes in the F1 generation, these genes are also required in F1 and later progeny for the silencing of germline-expressed genes (Ashe et al., 2012; Buckley et al., 2012). However, whereas the Argonaute NRDE-3 mediates F1 silencing in the soma, it is not required for germline silencing (Ashe et al., 2012; Buckley et al., 2012). Instead, another Argonaute named *heritable RNAi defective* (*hrde-1*, also known as *wago-9*; Gu et al., 2009; Table 1.1) was identified in genetic screens for RNAi inheritance-deficient mutants (Ashe et al., 2012; Buckley et al., 2012). HRDE-1/WAGO-9 shows germline-specific nuclear expression (Ashe et al., 2012; Buckley et al., 2012; Shirayama et al., 2012); it is required for the inheritance dsRNA-induced silencing in the F2 and F3 generations (Ashe et al., 2012; Buckley et al., 2012) and for epigenetic silencing initiated by other means (see Section 4.2.2). HRDE-1 was shown to interact with secondary siRNAs (Figure 1.1) and to promote the association of NRDE-2 with pre-mRNA and the induction of H3K9me3 at the target locus (Buckley et al., 2012). The biological role of HRDE-1 and other genes involved in exo- and endo-RNAi is discussed in Section 3.3.3.

3.1.9 Chromatin modulators of the RNAi response

As many components involved in the exogenous RNAi pathway appear to be rate limiting, it is possible that factors affecting their abundance may influence the efficiency of the RNAi response. Certain proteins with predicted chromatin-binding properties and some known regulators of transcription are particularly important in modulating the sensitivity of *C. elegans* to dsRNA-induced silencing.

Shortly after the discovery of the potency of dsRNA in inducing gene silencing, large-scale RNAi-based screens started to be used for identifying gene function in *C. elegans*. In one such approach, when pools of dsRNA molecules targeting several genes were used in an initial screen, it was discovered that downregulation of genes with predicted chromatin function—*zfp-1* and *gfl-1*—as well as H3K36 methyltransferase MES-4 (Rechtsteiner et al., 2010) and two genes encoding Polycomb-group-related proteins compromised the efficiency of other dsRNAs from the same pool (Dudley, Labbe, & Goldstein, 2002).

Conversely, mutations in other genes encoding chromatin regulators, such as *C. elegans* Retinoblastoma (Rb) homolog *lin-35*, were shown to promote the efficiency of RNAi (Lehner et al., 2006; Wang et al., 2005). These negative regulators of RNAi belong to the genetically defined group of SynMuv (Synthetic Multivulva) B genes that act redundantly with the SynMuv A and SynMuv C groups to negatively regulate the inappropriate initiation of Ras signaling and ectopic vulvae development (Ceol & Horvitz, 2004; Cui et al., 2006; Ferguson & Horvitz, 1989). The core protein complex formed by the SynMuv B gene products is a repressive chromatin complex (dREAM/Muv B) (Harrison, Ceol, Lu, & Horvitz, 2006). This complex, along with other associated proteins, was shown to inhibit germline-specific fate in the somatic tissues of *C. elegans* (Unhavaithaya et al., 2002; Wang et al., 2005). Interestingly, expression of a number of germline-enriched Argonaute proteins is enhanced in SynMuv B mutants (Grishok, Hoersch, & Sharp, 2008; Wu, Shi, Cui, Han, & Ruvkun, 2012), which suggests a potential reason for their enhanced susceptibility to RNAi (Figure 1.3).

There is a peculiar genetic relationship between the chromatin factors promoting RNAi, such as *zfp-1*, *gfl-1*, and *mes-4*, and the SynMuv B genes inhibiting RNAi efficiency. It was found that the ectopic vulvae formation seen in SynMuv B; SynMuv A double mutants is suppressed by mutations in a group of genes called SynMuv suppressors (Cui, Kim, & Han, 2006). Interestingly, *zfp-1*, *gfl-1*, and *mes-4* belong to the SynMuv suppressor group (Cui, Kim, & Han, 2006; Wang et al., 2005), and many other SynMuv suppressors have been found to promote RNAi (Cui, Kim, & Han, 2006). A mechanistic understanding of the relationship between SynMuv B genes and SynMuv suppressors is currently not clear. As ZFP-1 negatively modulates transcription (Mansisidor et al., 2011) and localizes to the promoters of SynMuv B genes (Mansisidor et al., 2011), it is possible that inhibition of *zfp-1* function leads to a diminished RNAi response due to enhanced expression of SynMuv B proteins and lower levels of Argonaute proteins (Figure 1.3). The effect of MES-4 in promoting Argonaute expression is likely to be direct (Gaydos, Rechtsteiner, Egelhofer, Carroll, & Strome, 2012; Kudron et al., 2013; Figure 1.3).

An interesting connection between miRNA function and regulation of RNAi efficiency has been discovered recently (Massirer, Perez, Mondol, & Pasquinelli, 2012). Although biologically relevant targets have not been

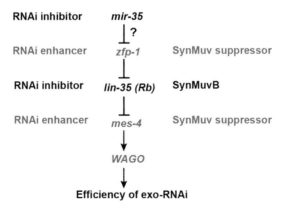

Figure 1.3 A hypothetical genetic pathway connecting chromatin-associated factors regulating exo-RNAi efficiency. Positive or negative connections between the components are based on the published data with the exception of negative regulation of *zfp-1* by *mir-35*, which is speculative (3′UTR of *zfp-1* contains a predicted *mir-35* binding site).

described for the abundant *mir-35* family expressed in the germline and embryo, a mutation eliminating a large number of *mir-35* family miRNAs, *mir-35–41(gk262)*, led to enhanced susceptibility to exogenous RNAi (Massirer et al., 2012). This effect was due to lower levels of LIN-35/Rb in *mir-35* mutants because ectopic expression of LIN-35 rescued the RNAi hypersensitivity phenotype. Notably, maternal expression of LIN-35 was shown to be sufficient for the restoration of the normal RNAi response in animals lacking zygotic *lin-35* function (Massirer et al., 2012). It would be interesting to determine how the *mir-35–41(gk262)* mutation leads to reduced LIN-35 Rb levels and whether negative regulation by *mir-35–41* of SynMuv suppressors with the potential to inhibit *lin-35*, such as ZFP-1, can explain the relationship between *mir-35–41* and LIN-35 (Figure 1.3).

3.2. Genomic and molecular features of endogenous RNAi

Initial studies of RNAi in *C. elegans* predicted its antiviral role, while miRNAs appeared as endogenous regulators of gene expression. Then, the discovery of an enhanced RNAi response in the RdRP mutant *rrf-3* suggested an endogenous role for the putative secondary siRNAs generated by this enzyme and a competition between the exo- and endo-RNAi pathways (Simmer et al., 2002). Indeed, endogenous siRNAs antisense to protein-coding genes were discovered shortly thereafter (Ambros et al., 2003) and an active investigation of endogenous RNAi in *C. elegans* began. By 2012, a significant body of knowledge had been generated, and a SnapShot summary comparing endogenous RNAi machinery and mechanisms across species has been published recently (Flamand & Duchaine, 2012).

3.2.1 Discovery of endo-siRNAs

Sequencing experiments aimed at identifying more miRNAs by Ambros and colleagues also yielded reads antisense to 551 different protein-coding genes as well as reads matching retrotransposons and a peculiar locus on the X chromosome termed "X cluster" (Ambros et al., 2003). At that time, a prevalence of guanosine in the $5'$ position of endo-siRNAs was noted (Ambros et al., 2003). Another sequencing report by the same group defined 1085 genes with matching antisense siRNAs (Lee et al., 2006). Initial analyses of the genetic requirements for endo-siRNA production determined that mutants compromised in exo-RNAi, such as *rrf-1*, *rde-3*, *mut-7*, and *mut-14*, as well as mutants with an enhanced RNAi response, such as *rrf-3* and *eri-1*, affected endo-siRNA accumulation (Lee et al., 2006). Interestingly, although both *rde-1* and *rde-4* are required for the exo-RNAi response, only *rde-4* appeared to have a role in endo-RNAi (Lee et al., 2006).

The development of the new high-throughput pyrophosphate sequencing method (Margulies et al., 2005) enabled a deeper analyses of short RNA populations in *C. elegans* and further classification of endo-siRNAs (Pak & Fire, 2007; Ruby et al., 2006). A prevalent length of either 21–22 or 26 nt and a 5′G in both kinds of siRNAs was reported (Ruby et al., 2006). The features of the 26-nt class of endo-siRNAs included a 5′monophosphate and a modified 3′terminus (Ruby et al., 2006), whereas the 22-nt class was underrepresented in libraries generated by a 5′monophosphate-ligation-dependent protocol and contained a 5′triphosphate like the secondary siRNAs generated by RdRPs during exo-RNAi (Pak & Fire, 2007; Ruby et al., 2006). Many unique cloned sequences matched splice junctions, which supported their origin from RdRP amplification using mature mRNA as a template; however, intron-matching reads were also found, which suggests the possibility of nuclear RdRP-dependent amplification (Pak & Fire, 2007; Ruby et al., 2006).

3.2.2 Distinction between WAGO- and CSR-1-associated 22G-endo-siRNAs

The application of the Illumina/Solexa deep sequencing platform stimulated the discovery of additional short RNAs in many organisms, including *C. elegans*. An earlier described class of endogenous 21–22 nt RNAs was significantly expanded and called 22G-RNAs (Gu et al., 2009). 22G-RNAs are most abundant in the germline, although somatic 22G have also been described. The two major classes of 22G-RNAs are defined by their interacting Argonaute proteins: the majority of 22G-RNAs antisense to protein-coding genes exist in complex with CSR-1 (Claycomb et al., 2009; Gu et al., 2009), while the WAGO family (required for the exogenous RNAi response) interacts with 22G-RNAs targeting transposons, pseudogenes, cryptic loci, and a few coding genes (Gu et al., 2009; Figures 3.4–3.6). 22G have features consistent with them being RdRP products, and they require RdRPs for their accumulation. RRF-1 and EGO-1 act redundantly in generating germline-enriched WAGO-22G-RNAs; somatic 22G-RNAs require RRF-1, and CSR-1-bound 22G are dependent only on EGO-1 (Claycomb et al., 2009; Gu et al., 2009). Dicer-related helicase DRH-3 and tudor-domain protein EKL-1 are required for the production of both classes of 22Gs and interact with RdRPs (Aoki et al., 2007; Gu et al., 2009), consistent with their characterization as the core RdRP module components (Thivierge et al., 2012; Figures 3.4–3.6). Interestingly, sequencing of short RNAs from non-null *drh-3* mutants with mutations in the putative helicase domain revealed the persistence of 22G-RNAs antisense to 3′UTR sequences, suggesting that RdRPs are recruited to these regions and that the helicase activity of DRH-3 is required for their access to more upstream target mRNA sequences (Gu et al., 2009). The abundance of WAGO-22G-RNAs is also dependent on RDE-3 and MUT-7 (Gu et al., 2009). In addition, the essential role of the MUT-16 protein in the formation and/or stability of both germline and somatic 22G-RNAs of the WAGO class was recently described (Zhang et al., 2011).

WAGO-22G-RNAs are more abundant than CSR-1 22G-RNAs, and they have a clear silencing effect on their targets, including suppression of transposon mobilization (Gu et al., 2009; Zhang et al., 2011). Although CSR-1 was shown to be associated with chromatin at regions targeted by its cofactor 22G-RNAs (Claycomb et al., 2009), only modest changes in gene expression were detected in *csr-1*(−) worms, which contrasts with their severe sterility phenotype (Claycomb et al., 2009; She, Xu, Fedotov, Kelly, & Maine, 2009; Yigit et al., 2006) and with the strong defects in chromosome segregation in animals treated with *csr-1* dsRNA (Claycomb et al., 2009; Yigit et al., 2006). Interestingly, the majority of genes corresponding to the antisense 22G-RNAs that are enriched in CSR-1 immunoprecipitation samples showed reduction in their expression in *csr-1*(−) sterile adults (Avgousti, Palani, Sherman, & Grishok, 2012; Claycomb et al., 2009). At the same time, analyses of 22G-RNA and mRNA abundance in *ego-1*(−) sterile worms revealed that ~300 germline genes were modestly upregulated in the absence of EGO-1-dependent 22G-RNAs (Maniar & Fire, 2011).

The abundance of CSR-1 22G-RNAs is controlled by the germline-specific nucleotidyltransferase protein CDE-1 (van Wolfswinkel et al., 2009). CDE-1 specifically uridylates CSR-1 22G-RNAs and interacts with RdRP EGO-1, but not with CSR-1 (van Wolfswinkel et al., 2009). Although the abundance of CSR-1 22G-RNAs is increased in *cde-1* mutants, these endo-siRNAs do not function properly, which is evident by the similarity of the chromosome segregation defects of *csr-1* and *cde-1* mutant worms (van Wolfswinkel et al., 2009). The function of the WAGO pathway in transposon silencing and exo-RNAi is also compromised in *cde-1* mutants, which suggests that CDE-1 is involved in proper separation of the 22G pathways (van Wolfswinkel et al., 2009).

Interestingly, a recent comparison of short RNA populations and their targets in four related nematodes revealed a higher than average conservation of genes targeted by CSR-1 22G-RNAs, whereas WAGO targets showed poor conservation and produced more siRNAs than conserved CSR-1 targets (Shi, Montgomery, Qi, & Ruvkun, 2013).

3.2.3 26G- and 22G-endo-siRNAs of the ERI pathway

The first enhanced RNAi (eri) gene, eri-1, was identified by mutant alleles with increased sensitivity to exogenous RNAi (Kennedy, Wang, & Ruvkun, 2004). The eri-1 gene encodes two protein isoforms, both of which contain a nucleic acid-binding SAP domain and a 3′-to-5′ exonuclease domain (Gabel & Ruvkun, 2008; Kennedy et al., 2004). Initially, the increased exo-RNAi response in eri-1 mutants was thought to be due to a suppressed degradation of siRNAs (Kennedy et al., 2004). Later, ERI-1 was recognized as a component of a specific endogenous RNAi pathway that is required for the production of certain endo-siRNAs and that competes with exo-RNAi (Duchaine et al., 2006; Figure 1.4). Also, ERI-1 was found among proteins coimmunoprecipitating with Dicer in an unbiased proteomic approach (Duchaine et al., 2006). Only the longer ERI-1b isoform functions in endo-RNAi, while the shorter ERI-1a isoform was shown to have a conserved role in 5.8S rRNA processing (Gabel & Ruvkun, 2008). Interestingly, the RdRP RRF-3, which also restricts exo-RNAi (Simmer et al., 2002), was also identified in the complex with Dicer (Duchaine et al., 2006), and two other eri genes, eri-3 and eri-5, encode Dicer complex components (Duchaine et al., 2006). Recent biochemical analyses revealed that the Dicer–RRF-3 core complex contains DRH-3 and Tudor domain protein ERI-5, which interacts with Dicer directly (Thivierge et al., 2012; Figure 1.4). The full list of ERI–Dicer-1 complex (ERIC) components identified by proteomic analyses also includes ERI-3, ERI-1, and RDE-4 (Thivierge et al., 2012).

The molecular function of ERIC is to produce a longer species of endogenous siRNAs called 26G-RNAs (Gent et al., 2010; Han et al., 2009; Vasale et al., 2010; Figure 1.4). Two classes of 26G-RNAs have been identified: Class I is present in purified sperm, while Class II is most abundant in oocytes and embryos and diminished in expression during postembryonic development (Han et al., 2009; Figure 1.4). Depletion of both types of 26G-RNAs affects the expression of their target genes, although the effect of depleting Class II 26G-RNAs persists much longer in development than the 26Gs themselves (Han et al., 2009). The fact that the ERI pathway also affects

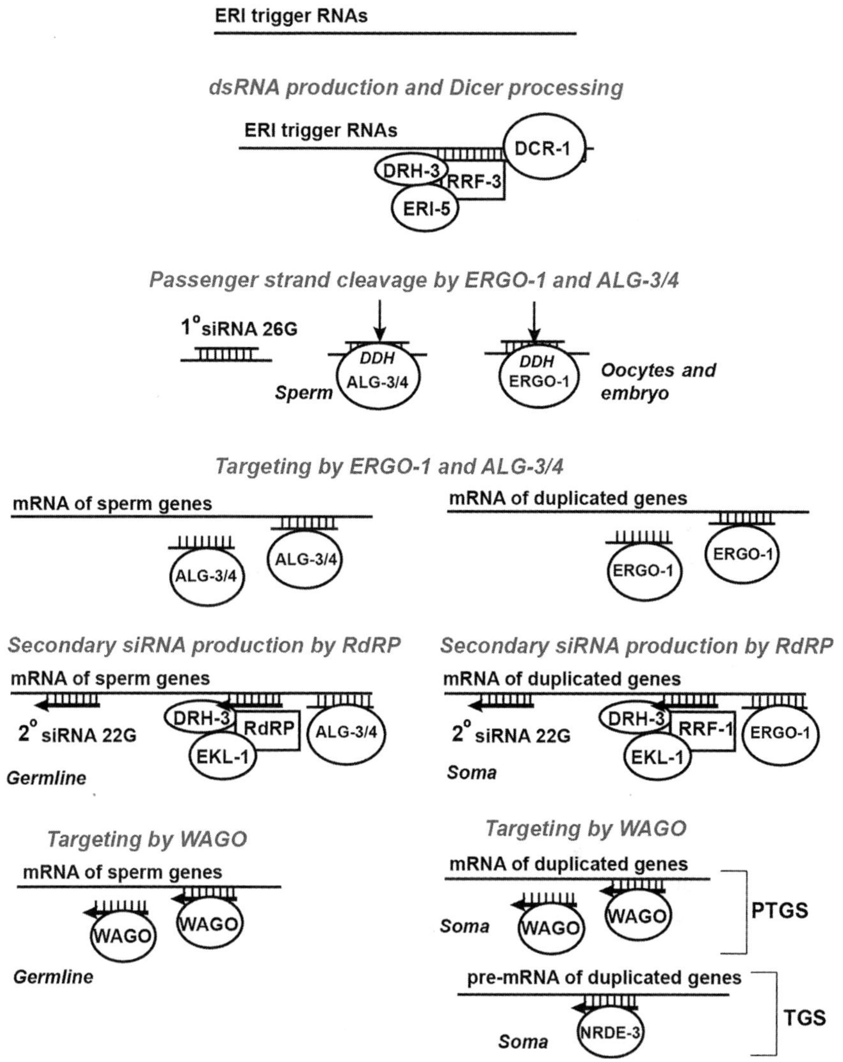

ERI trigger RNAs

dsRNA production and Dicer processing

ERI trigger RNAs

DRH-3 RRF-3 ERI-5 DCR-1

Passenger strand cleavage by ERGO-1 and ALG-3/4

$1°$siRNA 26G

DDH ALG-3/4 Sperm

DDH ERGO-1 Oocytes and embryo

Targeting by ERGO-1 and ALG-3/4

mRNA of sperm genes

ALG-3/4 ALG-3/4

mRNA of duplicated genes

ERGO-1 ERGO-1

Secondary siRNA production by RdRP

mRNA of sperm genes

$2°$ siRNA 22G DRH-3 RdRP EKL-1 ALG-3/4

Germline

Secondary siRNA production by RdRP

mRNA of duplicated genes

$2°$ siRNA 22G DRH-3 RRF-1 EKL-1 ERGO-1

Soma

Targeting by WAGO

mRNA of sperm genes

WAGO WAGO

Germline

Targeting by WAGO

mRNA of duplicated genes

Soma WAGO WAGO PTGS

pre-mRNA of duplicated genes

Soma NRDE-3 TGS

Figure 1.4 ERI-dependent endogenous RNAi pathways. Specific messages are selected as templates for RdRP RRF-3 by unknown mechanisms that include recognition of duplicated genes. dsRNA produced by the RRF-3 complex is cleaved by Dicer to produce 26G-duplex siRNAs where the passenger strand is ~19 nt. Two types of 26G-RNAs are recognized: sperm-specific 26G bound by ALG-3/4 and oocyte/embryo-specific 26G bound by ERGO-1. ERGO-1 and likely ALG-3/4 cleave the passenger strand in the 26G-siRNA duplex. mRNAs targeted by 26G-RNAs are used as templates for secondary 22G-RNA production; 22G-RNAs are incorporated into complexes with WAGO Argonautes which stimulate PTGS and TGS. Only a fraction of WAGO-bound 22G-RNAs is dependent on the ERI pathway.

production of some 22G-RNAs (Duchaine et al., 2006; Gu et al., 2009) suggested a connection between 26G- and 22G-RNAs. Indeed, analysis of the short RNAs found in complex with ERGO-1 Argonaute (Yigit et al., 2006) identified Class II 26G-RNAs (Vasale et al., 2010; Figure 1.4), and a subset of 22G-RNAs was found to be associated with 26G-RNA-producing loci (Vasale et al., 2010). Importantly, *rrf-1* and WAGO were required for the accumulation of these 22G-RNAs (Figure 1.4), but not for the accumulation of corresponding 26G-RNAs, which required *rrf-3* and *ergo-1*. Therefore, a two-step pathway involving two rounds of RdRP amplification (by RRF-3 and RRF-1) and two Argonaute types (primary-ERGO-1 and secondary-WAGO) emerged (Vasale et al., 2010; Figure 1.4). The accumulation of 22G-RNAs corresponding to Class I spermatogenesis-specific 26G-RNAs is dependent on ALG-3/4 Argonautes expressed in the germline region that undergoes spermatogenesis (Conine et al., 2010); therefore, the molecular function of ALG-3/4 is likely to be similar to that of ERGO-1 (Figure 1.4). An independent study based on deep sequencing of short RNAs from *rrf-3* mutant animals that identified somatic target genes regulated by the ERI pathway suggested a similar two-step model of 26G-RNA-dependent 22G-RNA production (Gent et al., 2010). Notably, 26G-RNAs, unlike 22G-RNAs, were found to be modified at their 3′ends (Ruby et al., 2006; Vasale et al., 2010). HENN-1 RNA methyltransferase was implicated in methylating the 3′ ends of the ERGO-1 class of 26G-RNAs, and this methylation had a stabilizing effect on these RNAs (Billi et al., 2012; Kamminga et al., 2012; Montgomery et al., 2012). In the absence of HENN-1, uridylation of 26Gs was frequently observed (Kamminga et al., 2012).

The somatic Argonaute protein NRDE-3, which binds to secondary siRNAs and translocates to the nucleus in an siRNA-dependent manner, has also been connected to the ERI pathway (Figure 1.4). NRDE-3 is localized to the nucleus in wild-type worms (Guang et al., 2008) but loses nuclear localization in mutants with defective ERI-dependent 22G production, including *eri-1*, *rde-4*, *ergo-1* (Guang et al., 2008), *eri-9*, *dcr-1(mg375Eri)* (Pavelec et al., 2009), and *eri-6/7* (Fischer et al., 2011). Moreover, silencing of a GFP-sensor regulated by the ERI-dependent 22G-siR-1 was shown to require NRDE-3 (Montgomery et al., 2012). Given the connection between 26G and 22G production, these results are consistent with the endogenous role of NRDE-3 in repression of 26G-RNA-target genes (Burkhart et al., 2011; Guang et al., 2008). Notably, initiation of GFP-sensor silencing by the NRDE-3-dependent 22G-siR-1 tolerated some degree of mismatching (Montgomery et al., 2012). *mut-2* and *mut-7* are also required for the nuclear localization of NRDE-3 (Guang et al., 2008), which suggests that they have a role in the ERI pathway. Indeed, *mut-16*, *mut-2*, and *mut-7* were shown to be essential for the accumulation of ERGO-1 26G-RNAs, and *mut-16* was partially involved in the production of secondary 22G-RNAs of the ALG-3/4 class (Zhang et al., 2011).

An initial survey of the loci targeted by ERGO-1-dependent 22G-RNAs revealed that they are preferentially localized to chromosome arms and enriched in gene duplications (Vasale et al., 2010; Figure 1.4). A more extensive investigation of ERGO-1 targets was conducted in a study of the ERI-6/7 helicase, which is encoded by two adjacent genes whose pre-mRNAs fuse in a rare *trans*-splicing event (Fischer, Butler, Pan, & Ruvkun, 2008). ERI-6/7 is homologous to helicases that act with short RNAs in plants and animals. Similar to ERGO-1, ERI-6/7 is required for the production of Class II 26G-RNAs abundant in oocytes and embryos and for that of their dependent 22G-RNAs, which are abundant in somatic tissues (Fischer et al., 2011). This temporal shift in the peaks of abundance of Class II 26Gs and their corresponding 22Gs suggests that RRF-3-dependent 26G production and RRF-1-dependent 22G generation are separated in time and that RRF-3 acts in the embryo, while RRF-1 acts postembryonically (Fischer et al., 2011). Interestingly, bioinformatic analyses of 78 genes matching *eri-6/7*-dependent siRNAs revealed groups of related genes that are poorly conserved in other species and have few introns (Fischer et al., 2011). These analyses are consistent with the ERGO-1 26G branch serving to silence recently duplicated or horizontally acquired genes (Fischer et al., 2011). This conclusion is further supported by a recent high-throughput sequencing study of the evolution of short RNAs using four related nematode species: *C. elegans*, *Caenorhabditis briggsae*, *Caenorhabditis remanei*, and *Caenorhabditis brenneri* (Shi et al., 2013). Although 26G-RNA sequences themselves are not conserved, it is remarkable that embryo-specific ERGO-1 26G-RNA-target genes showed only 2–12% conservation when considering the higher than average conservation of sperm-specific 26G-RNA-target genes (Shi et al., 2013). This study also detected conservation in the localization of ERGO-1 26G-RNA-producing loci to gene-poor chromosome arms (Shi et al., 2013).

The distinctive feature of 26G-RNA biogenesis is its Dicer dependence, which is in contrast to the Dicer-independent biogenesis of 22G-RNAs (Figure 1.4). The sequences of 26G-RNAs are not consistent with a processive generation by Dicer using long dsRNA substrates produced by

RRF-3 (Fischer et al., 2011; Vasale et al., 2010). Although 26G-RNAs are mostly 26 nt reads antisense to mRNAs in wild-type worms, stabilization of the passenger-strand short RNAs antisense to 26Gs was detected in *ergo-1* mutants, which is consistent with slicing of the passenger strand by ERGO-1 (Fischer et al., 2011). Interestingly, the passenger strand of the 26G duplex is ∼19 nt, such that ∼3–4 nt 5′ and 3′ overhangs of the longer 26G strand are present (Fischer et al., 2011; Figure 1.4).

The involvement of the helicase domain of Dicer in endo-siRNA production was demonstrated with the *dcr-1(mg375)* mutation, which was identified in genetic screens for ERI mutants (Pavelec et al., 2009). A more extensive study using transgenes with point mutations in the helicase domain, which were introduced into a *dcr-1(−)* null background, confirmed that this domain is essential for endo-siRNA production and demonstrated that it specifically affects 26G-RNA levels genome wide (Welker et al., 2010). An increased number of 26G reads was noted in *C. elegans* mutants of ADARs (Warf, Shepherd, Johnson, & Bass, 2012). The predominant 26G duplex RNA structures identified by this study contained 3 nt 3′ overhangs consistent with findings from studies of Dicer processing of long dsRNA (Welker et al., 2011; Figure 1.4). Therefore, the 3′ end of 26G-RNA results from Dicer cleavage, and it was suggested that the 3′–5′ nuclease ERI-1 is involved in the processing of the 3′ end of the 19-nt-long 26G passenger strand, which would result in more variability in the 5′ overhangs of the 26G duplex (Warf et al., 2012). The precise mechanism of 26G-RNA biogenesis involving the Dicer, ERI-1, and RRF-3 module remains to be elucidated.

Although RDE-1 and RDE-4 are equally required for exo-RNAi, RDE-4 has a much larger role in endo-RNAi. RDE-4 is required for 26G production (Vasale et al., 2010) and also appears to have a connection to EGO-1-dependent siRNAs (Maniar & Fire, 2011). However, the precise molecular function of RDE-4 in endo-RNAi is not clear. RDE-1 has a role in the production of most abundant somatic 22G-RNAs corresponding to the Y47H10A.5 gene, which is initiated by the *mir-243* miRNA that RDE-1 binds (Correa et al., 2010), and in the production of a few other 22G-RNAs. Notably, these RDE-1-dependent siRNAs are also depleted in *rde-10* and *rde-11* mutants, which otherwise affect few endo-siRNAs (Zhang et al., 2012).

3.2.4 Competition between ADARs and endogenous RNAi

There are two *C. elegans* genes encoding ADARs: *adr-1* and *adr-2* (Tonkin et al., 2002). Although only ADR-2 contains a catalytic domain and active editing function, ADR-1 modulates the activity of ADR-2 *in vivo* (Knight & Bass, 2002; Tonkin et al., 2002). It was shown earlier that somatic expression of repetitive transgenes was silenced in ADAR mutants in an *rde-1-* and *rde-4*-dependent manner (Knight & Bass, 2002). Therefore, a similar competition between endogenous RNAi and ADARs for dsRNA substrates can be expected. Indeed, deep sequencing analyses identified a number of low-to-moderate copy inverted repeat regions with a dramatic increase in short RNA reads in ADAR mutants; consistently, transcripts from such loci were found to be multiply edited (Wu, Lamm, & Fire, 2011). Interestingly, although a corresponding decrease in mRNA levels was often observed, histone messages remained unchanged in ADAR(−) animals despite dramatic increases in short RNAs corresponding to some histone loci (Wu et al., 2011).

The effect of ADARs on the biogenesis of short RNAs was examined in another study (Warf et al., 2012). Although the levels of many miRNAs were increased in the absence of ADARs, most of these effects were found to be indirect, likely due to decreased sequestering of pri-miRNAs by ADARs from Drosha processing; only a couple miRNAs were found to be edited by ADARs (Warf et al., 2012). Surprisingly, generation of endo-siRNAs, which were selected by a 5′monophosphate-dependent sequencing protocol, was predominantly suppressed in the ADAR mutants, with ∼40% of annotated loci producing fewer antisense siRNAs (Warf et al., 2012). These results are more consistent with the indirect effects of ADAR loss. Conversely, production of many Dicer-dependent 26G-antisense RNAs and their complementary 19 nt passenger strands was increased in the ADAR mutants, consistent with the competition between Dicer and ADARs for the dsRNA precursors of 26G-RNAs (Warf et al., 2012). As ADARs localize to the nucleus (Hundley, Krauchuk, & Bass, 2008), it is most plausible that the RRF-3-dependent synthesis of 26G-RNA precursors takes place there (Warf et al., 2012).

3.3. Biological functions of RNAi

3.3.1 Antiviral defense

The initiation of the RNAi response by dsRNA immediately suggested a natural antiviral role for this gene-silencing phenomenon (Fire et al., 1998). Moreover, the apparent lack of obvious developmental and physiological defects in *rde-1* and *rde-4* mutants further supported this idea (Tabara et al., 1999). This view of exogenous RNAi as a short RNA-based antiviral immunity was further confirmed by using transgenic animals expressing

Flock house virus, a plus-strand RNA animal nodavirus (Lu et al., 2005), as well as infections of *C. elegans* primary embryonic cell cultures with vesicular stomatitis virus, a rhabdovirus infecting insects and mammals (Schott, Cureton, Whelan, & Hunter, 2005; Wilkins et al., 2005).

An exciting development in studies of RNAi as an antiviral response came with the discovery of natural viruses infecting *C. elegans* and *C. briggsae* (Felix et al., 2011). Novel nodavirus-related RNA viruses, Orsay and Santeuil, were isolated from stably infected wild isolates of *C. elegans* and *C. briggsae*, respectively. Infection by these viruses resulted in clear morphological defects in the intestinal cells, which did not have a significant effect on animals, besides a lower rate of progeny production. A horizontal transmission of these viruses and a high specificity to *Caenorhabditis* species (*C. elegans* or *C. briggsae*) was described, as well as an efficient elimination of viruses by bleaching; no vertical transmission of the virus was detected in the progeny of bleach-treated parents. Consistent with an antiviral role of the exogenous RNAi pathway, individual *C. elegans* mutants defective for RNAi (*rde-1*, *rde-2*, *rde-4*, and *mut-7*) exhibited higher levels of viral RNA expression and more significant infection symptoms compared to the laboratory wild-type strain N2 (Felix et al., 2011). Also, deep sequencing of short RNA populations from infected animals revealed that 2% of all the unique sequences mapped to the viral RNA segments, and secondary 22G-RNAs were readily identified among these reads. Interestingly, a wild *C. elegans* isolate, JU 1580, which harbors the Orsay virus, was found to be resistant to RNAi targeting somatic but not germline genes (Felix et al., 2011). Other natural isolates of *C. elegans* also showed a variable sensitivity to somatic RNAi (Felix et al., 2011), and a wild-type isolate resistant to germline RNAi had been described earlier (Tijsterman, Okihara, et al., 2002). This natural variation in the RNAi responses could indicate a widespread coevolution of nematodes and nonlethal RNA viruses.

3.3.2 Silencing of transposons and repetitive elements
The first indication of the role of RNAi in genome surveillance came with the discovery that RNAi-deficient mutants often display increased rates of transposon mobilization in the germline, a mutator phenotype (Ketting et al., 1999; Sijen & Plasterk, 2003; Tabara et al., 1999; Tijsterman, Ketting, et al., 2002; Vastenhouw et al., 2003). The genome surveillance system in *C. elegans* comprises two 22G-RNA classes: the ERI/Dicer-dependent class and the ERI/Dicer-independent class (Gu et al., 2009; Figures 3.4 and 3.5). The 22G-RNAs of the ERI pathway target gene duplications and possibly horizontally transferred genes, as described in

Section 3.3, (Fischer et al., 2011; Vasale et al., 2010; Figure 1.4). Meanwhile, germline–enriched Dicer–independent 22G-RNAs bound by WAGO-1 and other WAGO Argonutes silence transposons, abberant transcripts, and cryptic loci (Gu et al., 2009; Figure 1.5). The generation of the WAGO

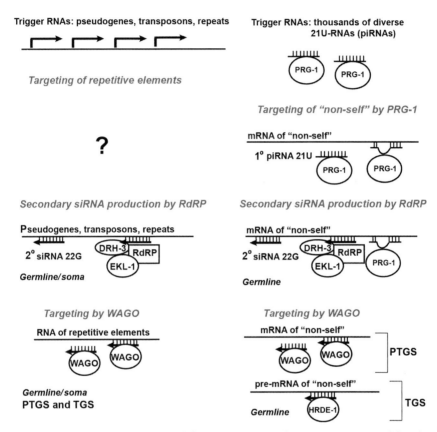

Figure 1.5 Repetitive regions and foreign genomic elements are targeted by the WAGO-22G-RNAs. Left: The initial steps promoting Dicer-independent 22G-RNA generation on repeats, pseudogenes, and transposons are poorly understood, but these 22G-RNAs are among the most abundant ones in *C. elegans*. The question mark refers to the possibility that some unknown Argonaute or a combination of Argonaute proteins bind to non-canonical primary siRNAs produced in a Dicer-independent manner and select mRNAs to be targeted by RdRP. It is also possible that some secondary siRNAs bound by WAGO are inherited by germline progenitor cells and stimulate RdRP activity on WAGO targets in the next generation; this system would then be Dicer and primary siRNA-independent, unlike the pathways shown in Figure 1.1, Figure 1.4 and Figure 1.5 (right panel). Right: 21U-RNAs bound by PRG-1 target foreign mRNA sequences (interaction does not require perfect complementarity) to promote RdRP-dependent production of WAGO-22G-RNAs. Notably, nuclear Argonaute HRDE-1 is responsible for initiating stable silencing of "non-self" sequences.

family 22G-RNAs is initiated in many cases by *C. elegans* piRNAs, and a combination of both types of short RNAs appears to have a more general role in recognizing foreign sequences distinct from endogenous genes (Figure 1.5; see Section 4.2.2). Importantly, the mutator proteins have been connected to the genome surveillance function of both the 26G–22G and piRNA-22G pathways (Gu et al., 2009; Zhang et al., 2011).

3.3.3 A summary of the global effects of 22G-RNAs on the expression of endogenous genes

The identification of endogenous short RNAs antisense to protein-coding genes poses a question concerning their relevance in gene regulation. To address this question, microarray analyses of gene misregulation in various mutant backgrounds have been performed (Asikainen, Storvik, Lakso, & Wong, 2007; Claycomb et al., 2009; Conine et al., 2010; Gent et al., 2009; Grishok et al., 2008; Gu et al., 2009; Lee et al., 2006; Maniar & Fire, 2011; van Wolfswinkel et al., 2009; Welker, Habig, & Bass, 2007). The general conclusion from these analyses is that the relationship between short RNAs and the gene loci from which they are generated, mostly by the action of RdRPs, is complex and specific to the endo-RNAi pathway to which these short RNAs belong. The WAGO family of 22G-RNAs has a strong negative effect on gene expression, which includes posttranscriptional, cotranscriptional, and transcriptional mechanisms in both the soma and the germline (Asikainen et al., 2007; Buckley et al., 2012; Conine et al., 2010; Duchaine et al., 2006; Fischer et al., 2011; Gent et al., 2010, 2009; Gu et al., 2009; Guang et al., 2008; Han et al., 2009; Lee et al., 2006; Pavelec et al., 2009; Vasale et al., 2010; Yigit et al., 2006; Zhang et al., 2011). However, as WAGO-22G-RNAs are largely dedicated to genome surveillance, they do not affect a large number of protein-coding genes and are most notably involved in spermatogenesis, with ALG-3/4 (Conine et al., 2010; Han et al., 2009), and maintenance of germline viability, with HRDE-1 (Buckley et al., 2012). It remains to be seen whether the germline mortality phenotype of the *hrde-1* mutant and some other *nrde* mutants is due to the gradually increased mobilization of transposons or due to the misregulation of protein-coding genes. On the other hand, the effects of the CSR-1 22G-RNAs on gene expression are modest (Claycomb et al., 2009; Maniar & Fire, 2011; van Wolfswinkel et al., 2009), though they represent the majority of short RNAs antisense to protein-coding genes, and the CSR-1 pathway mutants have the strongest developmental defects, described in Section 3.3.5. A recent study analyzed the effects of the RNAi pathways on changes in gene expression and chromatin modifications in adult animals that had experienced a stress-resistant and nonaging dauer stage (postdauer worms) compared with those that had experienced a normal life cycle (Hall, Chirn, Lau, & Sengupta, 2013). This work suggested a potential role for CSR-1 in somatic gene regulation, which was revealed by a significant increase in the level of H3K4 methylation in postdauer *csr-1* hypomorph mutants compared to wild-type postdauer worms (Hall et al., 2013).

In addition to gene silencing, some endo–siRNAs in *C. elegans* have a positive role in gene regulation (Figure 1.6). It was discovered recently that CSR-1 22G-RNAs have a role in promoting the processing of 3′UTRs of

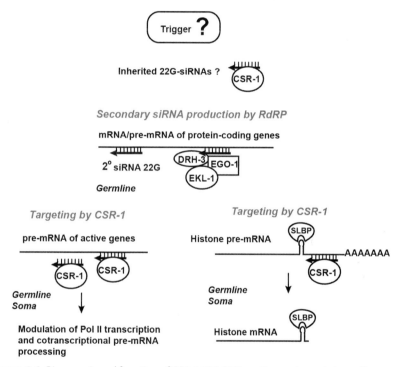

Figure 1.6 Biogenesis and function of CSR-1 22G-RNAs antisense to protein-coding genes. Triggers stimulating RdRP (EGO-1) recruitment to actively transcribed germline genes are not known at this time (question mark). The possibilities include currently unidentified primary siRNAs and their co-factor Argonautes or heritable secondary siRNAs produced by EGO-1 in the previous generation and deposited into germline precursor cells. CSR-1-bound 22G-RNAs modulate Pol II transcription of their targets and appear to positively regulate transcription in the germline. In addition to stimulating histone gene transcription, CSR-1 siRNAs promote histone mRNA biogenesis, possibly by guiding direct cleavage by CSR-1 after the conserved stem-loop. The stem loop in histone pre-mRNA and mRNA is recognized by the conserved stem-loop-binding protein (SLBP), which is required for proper histone mRNA biogenesis, nuclear export, and translation.

histone pre-mRNAs and that the deficiency in this process significantly contributes to the lethal and sterile phenotypes of the CSR-1 pathway mutants (Avgousti et al., 2012). Whether the majority of CSR-1 22G-RNAs targeting other genes also have an effect on pre-mRNA processing is not clear but presents an intriguing possibility. Consistent with this idea, a recent study of the evolutionary history of proteins in 85 genomes, which aimed to find genes with phylogenetic profiles similar to that of *C. elegans* RNAi-related factors, revealed a deep connection between RNAi and splicing (Tabach et al., 2013). Moreover, a significant number of splicing factors identified in this manner were shown to affect RNAi-dependent silencing of transgenes (Tabach et al., 2013).

3.3.4 Endogenous RNAi and adaptation to environment

Although the role of RDE-4 in exo-RNAi is understood relatively well (see Section 3.1.1), its contribution to endogenous RNAi pathways is defined less clearly. RDE-4 functions with RDE-1 in the rare cases when this Argonaute is involved in gene regulation (Correa et al., 2010; Gu et al., 2009), and it also coimmunoprecipitates with ERI-1 and ERI-5 (Thivierge et al., 2012) and is required for 26G production (Vasale et al., 2010). Moreover, *rde-4* is required for the maximum accumulation of 22G-RNAs produced by the germline RdRP EGO-1 (Maniar & Fire, 2011).

Microarray studies performed on *rde-4*-null worms identified a connection to regulation of stress response (Grishok et al., 2008; Mansisidor et al., 2011; Welker et al., 2007). Consistently, *rde-4* mutants have a decreased life span (Mansisidor et al., 2011; Welker et al., 2007) and an increased sensitivity to oxidative stress, pathogens (Mansisidor et al., 2011), and elevated temperatures (Blanchard et al., 2011). The signature of gene misregulation in *rde-4* mutant L1–L2 larvae significantly overlaps with that of *zfp-1* loss-of-function mutants (Grishok et al., 2008). ZFP-1 is a chromatin-binding protein enriched at promoters of highly expressed genes (Avgousti et al., 2012; Mansisidor et al., 2011). Intriguingly, only gene sets upregulated in the *zfp-1* and *rde-4* mutant worms (but not the downregulated gene sets) are enriched in endo-siRNA targets (Grishok et al., 2008). Although the majority of genes with increased expression in the *zfp-1* and *rde-4* mutants match CSR-1 22G-RNAs, the *pdk-1* gene, which encodes a conserved insulin signaling kinase whose upregulation is responsible for the shortened life span and the increased sensitivity to oxidative stress of the *zfp-1* and *rde-4* mutants, is a WAGO-22G target (Mansisidor et al., 2011). WAGO-22G-RNAs largely match the *pdk-1* promoter, which contains repeats that appear to support the production of dsRNA (Mansisidor et al., 2011). Interestingly, many genes upregulated in *rde-4* mutants contain repeats in their promoters (Mansisidor et al., 2011). It is possible that the production of specific 22G-RNAs from repetitive elements that match genes with adaptive function is naturally selected to "fine-tune" their expression. It is equally possible that modulation of genes without repeats in their promoters by CSR-1 22G-RNAs is also subject to natural selection.

3.3.5 Developmental roles of the CSR-1 22G-RNA pathway

The severe germline defects in mutants of RdRP EGO-1 provided the first indication of the relevance of RNAi to the regulation of developmental processes (Smardon et al., 2000). The *ego-1*(−) germline abnormalities included (1) a reduced proliferation of germ cells in the mitotic zone, a phenotype observed when GLP-1/Notch signaling is defective; (2) a premature entry of germ cells into meiosis and a slow progression through early meiotic prophase; and (3) the presence of oocytes with abnormal morphology and apparent chromatin condensation defects (Smardon et al., 2000; Vought, Ohmachi, Lee, & Maine, 2005). Moreover, embryos produced by the *ego-1* mutant mothers were noted to arrest in development after only a few cell divisions (Smardon et al., 2000; Vought et al., 2005). The *ego-1* activity was also implicated in the accumulation of H3K9 methylation on the unpaired and heterochromatin-rich single X chromosome in meiotic male germline and unpaired extrachromosomal fragments in hermaphrodites (Maine et al., 2005). In addition, abnormalities in the nuclear pore complex morphology in germ nuclei and in the germ cell-specific P-granule size and distribution were observed in *ego-1*(−) animals (Vought et al., 2005). In a survey of RNAi-related mutants, *csr-1*, *drh-3*, and *ekl-1* were identified as having phenotypes similar to *ego-1*(−) (She et al., 2009). Notably, Dicer mutants did not share these phenotypes (Maine et al., 2005). Several other studies reported chromosomal abnormalities in the oocytes and chromosome segregation defects in the embryos of *drh-3* (Claycomb et al., 2009; Duchaine et al., 2006; Nakamura et al., 2007), *csr-1* (Claycomb et al., 2009; Yigit et al., 2006), *ego-1*, *ekl-1* (Claycomb et al., 2009), and *cde-1* mutants (van Wolfswinkel et al., 2009). Further, P-granule defects were noted in the *csr-1* mutant (Claycomb et al., 2009) and uncovered in animals treated with *drh-3*, *csr-1*, and *ego-1* dsRNA in an unbiased RNAi screen for genes affecting P-granule function (Updike & Strome, 2009). These phenotypic analyses match well with the genomic studies separating CSR-1 and WAGO-22G pathways (see Section 3.2.2).

The cloning of short RNAs from the CSR-1 immunoprecipitates discovered its association with 22G-RNAs antisense to protein-coding genes (Claycomb et al., 2009; Figure 1.6). It was proposed that all these 22G-RNAs act together to specify chromatin features and architecture of holocentric *C. elegans* chromosomes (Claycomb et al., 2009). However, it was discovered recently that canonical histone proteins are severely depleted in the *csr-1*, *drh-3*, and *ego-1* mutant animals and that histone mRNAs are not processed properly in these mutants (Avgousti et al., 2012; Figure 1.6). Moreover, the overexpression of histone locus containing one copy of each core histone gene from a transgenic array significantly rescued the embryo lethality induced by the depletion of *csr-1* and *ego-1* by RNAi (Avgousti et al., 2012). These results suggest that histone depletion is a major contributing factor to the embryonic lethality seen in the CSR-1 pathway mutants and may also help explain many other germline abnormalities resulting from the defects in chromatin condensation. Despite this, P-granule defects in the RNAi mutants are not due to histone depletion (Avgousti et al., 2012).

Histone pre-mRNAs do not contain introns, but they require a specific 3′end processing event: a cleavage after the conserved stem-loop (reviewed in Marzluff, Wagner, & Duronio, 2008). In most organisms, there is a dedicated U7 snRNA, which base-pairs with a conserved sequence downstream of the stem-loop and recruits the cleavage complex (reviewed in Marzluff et al., 2008). The U7 snRNA and the conservation of its complementary downstream sequence are missing in nematodes (Davila Lopez & Samuelsson, 2008), but there are EGO-1-dependent 22G-RNAs complementary to the region downstream of the stem loop (Avgousti et al., 2012; Figure 1.6). As CSR-1 binds to histone mRNA and pre-mRNA (Avgousti et al., 2012) and is proficient in secondary siRNA-guided endonucleolytic cleavage (Aoki et al., 2007), it is most plausible that CSR-1 is an endonuclease which is responsible for the processing of histone pre-mRNA in *C. elegans* (Figure 1.6). However, it is also possible that CSR-1 acts to recruit the processing complex.

Interestingly, the germline function of CSR-1 also includes an interaction with the PUF (Pumilio/FBF) protein FBF-1 in the distal germline and the repression of translational elongation of FBF-1 target mRNAs (Friend et al., 2012). It is not clear whether this role of CSR-1 requires 22G-RNAs as the relevant phenotypes were less robust in the RdRP complex mutants: *ego-1*, *drh-3*, and *ekl-1* (Friend et al., 2012). Another example of the developmental role of the CSR-1 pathway genes is their requirement, redundantly with the KSR-1 scaffolding protein, in the specification of the excretory duct, worm's renal system, which occurs during embryogenesis (Rocheleau et al., 2008). Which 22G-RNA-target genes are relevant for this process and whether 22G-RNAs act positively or negatively to regulate them are not known. As there are two isoforms of the CSR-1 protein and one of them is expressed in somatic tissues (Claycomb et al., 2009), additional contributions of CSR-1 and 22G-RNAs to gene regulation during postembryonic development are likely to be discovered.

3.3.6 Cooperation between RNAi and Rb

Although the exo-RNAi pathway mutants display few obvious phenotypes, it is possible that a more careful examination could reveal the contribution of this pathway to the development and/or fitness of nematodes. The importance of *rde-4* for normal longevity and stress resistance has been discussed earlier (see Section 3.3.4). In addition, genetic studies that were conducted before the classification of endo-siRNAs revealed the redundant roles of the exo-RNAi factors and Rb in the negative regulation of nuclear divisions in the intestinal cells at the early L2 stage (Grishok & Sharp, 2005; Ouellet & Roy, 2007). Although the RNAi mutants did not exhibit an increase in nuclei number, double mutant combinations of the presumptive *lin-35* (Rb) null alleles with *rde-4*, *rde-1*, or *mut-7* nulls significantly enhanced the supernumerary nuclei phenotype of *lin-35*(−) (Grishok & Sharp, 2005; Ouellet & Roy, 2007).

The nuclear divisions in the intestine are very sensitive to cyclin E dosage (Grishok & Sharp, 2005), and the supernumerary phenotype is suppressed by mutations in cyclin E (Grishok & Sharp, 2005). Cyclin E mRNA expression is elevated in *lin-35*(−) worms (Grishok & Sharp, 2005; Ouellet & Roy, 2007), and its level is further enhanced in the *lin-35; rde-4* double mutant strain (Grishok & Sharp, 2005). The cyclin E gene (*cye-1* in *C. elegans*) is a known conserved target of transcriptional repression by Rb, and an antisense RNA overlapping exon 1 of *cye-1* was detected (Grishok & Sharp, 2005). Therefore, regulation of this gene by Rb and RNAi is likely to be direct. The possibility of dsRNA production at the *cye-1* locus is consistent with the involvement of RDE-1 in the regulation of nuclear divisions. Additional phenotypes associated with the combination of *lin-35* and *rde-4* null mutations include egg laying defects, extra vulva protrusions, and gonad migration defects (Grishok & Sharp, 2005). As *rde-4* appears to have a role in multiple RNAi pathways (exo-RNAi, ERI, CSR-1), it remains to be seen which other RNAi mutants will have similar phenotypes when combined with *lin-35*(−) and misregulation of which target genes is responsible for the phenotypes.

3.3.7 The role of ALG-3/4 endo-siRNAs in sperm development

The temperature-sensitive sterility and "High Incidence of Males" (Him) phenotypes, presumably associated with defects in sperm development, were reported in early studies of the mutator and Eri genes (Duchaine et al., 2006; Kennedy et al., 2004; Ketting et al., 1999; Simmer et al., 2002). These signature phenotypes were also found in the compound mutant strains of WAGO Argonautes (Gu et al., 2009; Yigit et al., 2006) and in the viable loss-of-function *drh-3* mutants (Gu et al., 2009). Consistently, the microarray analysis of L4-stage *rrf-3* and *eri-1* mutants identified the misregulation of genes required for sperm function (Asikainen et al., 2007).

Genetic analyses of the Eri mutants separated the enhanced RNAi sensitivity phenotype from the sperm-related sterility and Him phenotypes such that *eri-1*, *eri-3*, *eri-5*, *rrf-3*, and *dcr-1(mg375Eri)* mutant animals exhibited both phenotypes, whereas *ergo-1* and *eri-9* mutants did not show sterility and Him phenotypes but had a reduction in brood size at 25 °C (Pavelec et al., 2009). Also, *eri-6/7* mutants were not overtly sterile at elevated temperatures (Fischer et al., 2008) with some reduction in brood size (Fischer et al., 2011). These phenotypic analyses fit well with the genomic studies discussed earlier (see Section 3.2.3) that identified two classes of 26G-RNAs, only one of which was detected in sperm (Han et al., 2009; Figure 1.4). Therefore, the upstream Eri genes that correspond to the RRF-3 RdRP module (Thivierge et al., 2012) are required for both types of 26G-RNAs (Fischer et al., 2011; Han et al., 2009; Vasale et al., 2010) and are expected to have pleiotropic phenotypes, while other factors, such as the redundant Argonaute proteins ALG-3 and ALG-4, are required for the production of only sperm-specific 26G- and 22G-RNAs (Conine et al., 2010; Han et al., 2009; Figure 1.4). The phenotypes of the MAGO12 mutant strain are consistent with its role with ALG-3/4 26G-RNAs (Conine et al., 2010; Yigit et al., 2006). Since the Mutator mutants are Him, it could be expected that they act with sperm-specific 22G-RNAs. However, Mutators are not required for the production of sperm-specific 26G-RNAs, only partially affect ALG-3/4-dependent 22G levels, and are only partially rescued by male mating (Zhang et al., 2011).

The detailed analyses of 26G-related sperm development phenotypes were conducted in several studies (Conine et al., 2010; Gent et al., 2009; Pavelec et al., 2009). First, it was established that temperature-dependent sterility was associated with defects in sperm, not oocytes (Conine et al., 2010; Gent et al., 2009; Pavelec et al., 2009). The X-chromosome nondisjunction leading to the increased incidence of males (Him phenotype) was also connected to spermatogenesis (Gent et al., 2009). Although early events in germ cell development appeared to be unaffected in the Eri mutants (Pavelec et al., 2009), multiple abnormalities in spermatogenesis and spermiogenesis were described, including cell division defects (Conine et al., 2010; Gent et al., 2009; Pavelec et al., 2009), defects in spermatid activation (which affects sperm motility) (Conine et al., 2010; Pavelec et al., 2009), as well as sperm-related paternal effects leading to abnormal mucrotubule structures in early embryos and resulting embryonic lethality (Gent et al., 2009).

Although misregulation of many sperm-specific genes was reported in the mutants affecting ALG-3/4 endo-siRNAs (Conine et al., 2010; Gent et al., 2009; Han et al., 2009; Pavelec et al., 2009), the connection between the misregulated genes and the phenotypes is not clear. Understanding how regulation of specific genes by endogenous RNAi contributes to development and fitness is the next challenge for the ALG-3/4 class of short RNAs and for endo-siRNAs antisense to protein-coding genes in general.

4. 21U-RNAs, *C. elegans* piRNAs

piRNAs are the third major class of endogenous short RNAs, which is specific to animals, unlike miRNAs and endo-siRNAs that exist in both animals and plants. piRNAs interact with the PIWI subfamily of the Argonaute proteins. They were initially described in *Drosophila*, where they play a major role in transposon silencing utilizing both posttranscriptional and transcriptional mechanisms (reviewed in Ishizu et al., 2012; Sabin, Delas, & Hannon, 2013). Another distinct feature of piRNAs is their preferential expression in germ cells. Sterility phenotypes of animals lacking piRNAs are due to the secondary effects of transposon mobilization in *Drosophila* (reviewed in Khurana & Theurkauf, 2010), but it is not clear whether this is also true for nematodes and mammals and whether piRNAs participate in regulation of endogenous genes.

4.1. Biogenesis of 21U-RNAs

21U-RNAs were discovered as a separate class of short RNAs with unique features that included $5'$ uridine, 21-nt length, an apparent $3'$end modification, and the existence of two upstream genomic motifs (Ruby et al., 2006). One motif with the CTGTTTCA consensus is located at an invariant position preceding the 21U sequence and a smaller YRNT motif is present immediately upstream and ends with the 5'U of the 21U-RNA (Ruby et al., 2006; Figure 1.7). The 21U-RNA sequences themselves were found to be very diverse and the 21U-RNA-producing loci mapped to two broad regions on chromosome IV (Ruby et al., 2006). Although piRNAs in other species are longer (~26 nt), 21U-RNAs were recognized as functional equivalents of piRNAs due to their association with the *C. elegans* PIWI homolog, PRG-1 (Figure 1.5), and the dramatic reduction of 21U-RNA expression in *prg-1*(−) animals (Batista et al., 2008; Das et al., 2008; Wang & Reinke, 2008). Also, production of 21U-RNAs was shown to be Dicer-independent (Batista et al., 2008; Das et al., 2008). At the same time, the significance of the upstream motif in 21U-RNA expression was suggested by the higher abundance of 21U-RNAs with upstream motifs that

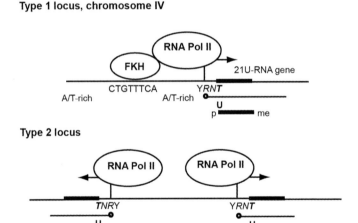

Figure 1.7 Biogenesis of 21U-RNAs (piRNAs). Each 21U-RNA is produced from a separate transcript defined by the YRNT sequence, where R (-2 relative to U) is the first nucleotide in the nascent 21U-RNA precursor. Two types of the 21U-RNA loci have been described: type 1 loci are prevalent on chromosome IV and contain nucleosome-depleted Pol II promoters with the CTGTTTCA DNA cis-element recognized by Forkhead transcription factors (FKH), and type 2 loci are present genome-wide and correspond to transcription start sites of protein-coding genes and other transcripts located in close proximity to promoters of protein-coding genes.

match closer to the consensus sequence (Batista et al., 2008). 21U-RNAs appear to be restricted to the germ tissue (Batista et al., 2008; Das et al., 2008; Wang & Reinke, 2008) and are expressed in both male and female germline (Batista et al., 2008; Das et al., 2008). Moreover, separate groups of male-enriched and female-enriched 21U-RNAs were recognized recently (Billi et al., 2013), and this separation also exists in other nematodes (Shi et al., 2013). Although genomic distributions of male-enriched and female-enriched 21U-RNAs are similar in *C. elegans*, these sub-groups are produced from different chromosomes in *C. briggsae* suggesting their different evolutionary history (Shi et al., 2013).

In other animals, numerous piRNAs are generated from one long precursor transcript (reviewed in (Ishizu et al., 2012)). However, the conserved motif associated with each individual 21U-RNA strongly suggested their autonomous expression in *C. elegans* (Ruby et al., 2006). Indeed, the upstream region is depleted of nucleosomes, which is a prominent feature of the promoters, and Pol II enrichment was only detected there in animals with germline tissue (Cecere, Zheng, Mansisidor, Klymko, & Grishok, 2012). Moreover, the CTGTTTCA requirement for 21U-RNA expression was demonstrated in experiments where this motif was deleted (Cecere et al., 2012) or scrambled (Billi et al., 2013; Figure 1.7). Furthermore, the sufficiency of the core motif for 21U-RNA expression was shown with single copy minimal expression cassettes integrated at genomic sites far away from the 21U clusters (Billi et al., 2013). A specific recognition of the CTGTTTCA consensus sequence by the Forkhead family of transcription factors was demonstrated *in vitro*, and binding of a Forkhead protein to the upstream regions of several 21U-RNAs was shown *in vivo* (Cecere et al., 2012; Figure 1.7). A decrease in 21U-RNA expression was correlated with the depletion of Forkhead proteins, many of which appear to act redundantly in promoting 21U-RNA expression (Cecere et al., 2012). Notably, the consensus **C**TGTTTCA sequence is enriched at male-specific 21U-RNAs promoters, whereas female-specific 21U-RNAs do not show a bias for C at the first position of the 8-mer (Billi et al., 2013). The 5' cytidine was shown to be important for the biased expression of 21U-RNAs during spermatogenesis, although oogenesis-biased expression of 21U-RNAs cannot be explained by the differences in the 8-mer sequence (Billi et al., 2013). Although there is a notable Pol II enrichment at the 21U-RNA promoters (Cecere et al., 2012), it is orders of magnitude lower compared to the promoters of actively expressed germline genes (Cecere et al., 2012). Also, mRNA expression of many Forkhead protein-encoding genes shows germline enrichment, but the corresponding protein expression appears to be very low (Cecere et al., 2012). The 5′ nucleotide in the Forkhead-specific DNA consensus sequence does not have a role in the interaction with the DNA-recognizing alpha helix of the Forkhead transcription factors, but it may provide a stabilizing contact with additional loop regions of the proteins (reviewed in (Obsil & Obsilova, 2011)) and therefore may determine the preferential binding of some male germline-enriched Forkhead family members to the **CTGTTTCA** consensus.

The 21U 8-mer upstream consensus sequence could not be detected as part of the precursor RNA by 5′-RACE experiments (Cecere et al., 2012) and by studies utilizing deep sequencing methods for the genome-wide annotation of long and short 5′ capped transcripts (Gu et al., 2012). Instead, the 5′ends of the 21U-RNA precursors map precisely to the purine (R) in the YRNT motif such that the nascent transcripts have two additional nucleotides preceding the first U of the mature 21U-RNA (Cecere et al., 2012; Gu, Lee, et al., 2012; Figure 1.7). The majority of 21U-RNA precursors appear to be ∼26 nt long (Gu, Lee, et al., 2012), although longer ones also exist (Cecere et al., 2012; Gu, Lee, et al., 2012). Therefore, the biogenesis of 21U-RNAs must include the removal of the cap and two nucleotides at the 5′end (Cecere et al., 2012; Gu, Lee, et al., 2012), and trimming and methylation at the 3′end (Billi, Alessi, et al., 2012; Kamminga et al., 2012; Montgomery et al., 2012; Figure 1.7).

Interestingly, additional *prg-1*-dependent 21U-RNAs produced from many chromosomes were recognized in later studies (Gu, Lee, et al., 2012; Kato et al., 2011). These 21U-RNAs are not associated with the conserved 8-mer motif (Gu, Lee, et al., 2012; Kato et al., 2011; Figure 1.7). However, the YRNN motif, where R represents the first transcribed nucleotide, was shown to be the general feature of Pol II transcription start sites in *C. elegans*, and Gu and co-authors found that the sites containing U as the third transcribed nucleotide (YRNT) produced 21U-RNAs (Gu, Lee, et al., 2012; Figure 1.7). In this study, 42% of the sense reads for ∼26-nt 21U-RNA precursors corresponded to the 5′ends of longer mRNA transcripts (Gu, Lee, et al., 2012). Also, 21U-RNAs of this second, 8-mer-independent, type (Type 2) were generally enriched within 1000 bp upstream of the 5′ends of transcript annotations (Gu, Lee, et al., 2012). Currently, it is not clear how transcription of these Type 2 21U-RNAs is related to their neighboring or overlapping protein-coding genes and whether the ∼26 nt 21U-RNA precursors arise from promoter-proximal Pol II pausing as has been suggested (Gu, Lee, et al., 2012).

4.2. Biological functions of 21U-RNAs

4.2.1 Function in fertility

The phenotype resulting from *prg-1(RNAi)* treatment was noted before the discovery of 21U-RNAs (Cox et al., 1998). It included a reduction in the size of the mitotic proliferation zone in RNAi-treated animals and a decrease in the number of progeny produced (Cox et al., 1998). In later studies, a decrease in both mitotic and meiotic germ nuclei was reported, although the change in the mitotic zone was more pronounced (Batista et al., 2008). Interestingly, this defect in *prg-1* mutant worms, as well as the decrease in 21U-RNA abundance, is not temperature-dependent, but the sterility of *prg-1(−)* animals is (Batista et al., 2008; Wang & Reinke, 2008; Yigit et al., 2006). It appears that the gametes in *prg-1* mutants are affected by a temperature-sensitive process (Batista et al., 2008; Wang & Reinke, 2008), and a defect in sperm activation may be one of the abnormalities leading to sterility (Wang & Reinke, 2008). Microarray profiling of dissected gonads from wild-type and *prg-1* mutant males detected a decrease in the expression of spermatogenesis genes (Wang & Reinke, 2008), but no dramatic changes in gene expression were detected using whole animals (Batista et al., 2008). A notable exception was the reversion of Tc3 transposon silencing and an increase in Tc3 mobilization in *prg-1* mutants (Batista et al., 2008; Das et al., 2008). Tc3 silencing was shown to be associated with an accumulation of endo-siRNAs that required *prg-1* function (Batista et al., 2008; Das et al., 2008), and a more extensive connection between *prg-1* and endo-siRNAs has been described recently (see Section 4.2.2). Importantly, a careful examination of *prg-1* mutant worms revealed their progressive sterility in generations due to germline mortality (E. Miska and S. Ahmed, personal communication). This phenotype bears some similarity to that of *hrde-1* mutant worms (discussed in Section 3.3.3) and could strengthen the biological significance of the connection between *C. elegans* piRNAs and endo-siRNAs.

4.2.2 Connection between 21U-RNAs, 22G-RNA production, and stable silencing of foreign DNA

Although the endogenous target genes regulated by 21U-RNAs were not immediately obvious, the use of single copy GFP-sensor transgenes confirmed the competence of 21U-RNAs in inducing gene silencing (Bagijn et al., 2012; Lee et al., 2012; Luteijn et al., 2012). Perfect complementarity between the 21U-RNA and its target site was not required for the initiation of silencing, as two mismatches were tolerated (Bagijn et al., 2012;

Lee et al., 2012). Consistently, PRG-1 proteins with mutations in the catalytic site were efficient at inducing silencing (Bagijn et al., 2012; Lee et al., 2012). Interestingly, an abundant production of antisense 22G-RNAs complementary to sequences 5' upstream of the 21U-RNA GFP-sensor target site was detected (Bagijn et al., 2012; Lee et al., 2012; Luteijn et al., 2012). This 22G generation and the process of the reporter silencing itself were shown to be dependent on PRG-1 and the RNA methyltransferase HENN-1, which are involved in 21U-RNA biogenesis, on nuclear RNAi pathway genes *nrde-1*, *nrde-2*, *nrde-4*, and *hrde-1/wago-9*, and on other components required for secondary siRNA production: RdRPs EGO-1 and RRf-1, RdRP complex component DRH-3, and mutator genes *mut-7*, *rde-2/mut-8*, *mut-14*, and *rde-3/mut-2* (Ashe et al., 2012; Bagijn et al., 2012; Lee et al., 2012; Luteijn et al., 2012). Importantly, PRG-1 was shown to be required for the initiation but not for the maintenance of reporter silencing, which was very stable across generations (Bagijn et al., 2012; Lee et al., 2012; Luteijn et al., 2012), much more stable than previously described exo-RNAi-induced heritable silencing (see Section 3.1.8). Consistently, two SET domain-containing predicted histone methyltransferase genes and the gene *hpl-2*, which encodes a *C. elegans* homolog of Heterochromatin Protein 1(HP1), were implicated in 21U-RNA sensor silencing but not in 22G-RNA production (Ashe et al., 2012), and H3K9 methylation was shown to be enriched at the silenced sensor sequences (Luteijn et al., 2012).

The phenomenon of 21U-RNA-induced transgene silencing discussed above has also been described for single copy transgenes not designed to be PRG-1-dependent silencing reporters (Shirayama et al., 2012). This type of silencing could be transmitted in a dominant fashion through crosses to affect other single copy transgenes that were otherwise stably expressed (Shirayama et al., 2012); such transmission in crosses was also a feature of the 21U reporter silencing (Luteijn et al., 2012). Other features of this phenomenon included dependence on *prg-1* for the initiation of silencing, but not for its maintenance, and dependence on *rde-3*, *mut-7*, *hpl-2*, and *wago-9/hrde-1* (Shirayama et al., 2012), in agreement with the studies described above. In addition, Polycomb and Trithorax complex-related factors were implicated in silencing, as well as the cytoplasmic WAGO-1 and nuclear WAGO-10 (Shirayama et al., 2012). The silencing mechanism appeared to combine both posttranscriptional and transcriptional regulation and was associated with an increase in H3K9 methylation (Shirayama et al., 2012). An interesting feature of this system is that efficient 22G-RNA production was detected near sites with partial 21U-RNA complementarity, which were not designed but identified later (Shirayama et al., 2012). The model suggested by these studies postulates that the diversity of 21U-RNAs makes them analogous to antibodies in the immune system, the diversity of which allows for the identification of any possible foreign targets distinct from "self" (Shirayama et al., 2012; Figure 1.5). The important question is how the organism defines "self" and protects it from silencing. It has been noted that transgene sequences corresponding to the endogenous genes targeted by the CSR-1 system accumulate fewer 21U-RNA-dependent 22G-RNAs. Therefore, it was suggested that CSR-1 22G-RNAs may play this antisilencing role (Luteijn et al., 2012; Shirayama et al., 2012). Curiously, there are experimental data consistent with the possibility that anti-silencing also requires *prg-1* activity (Luteijn et al., 2012). In any case, it is very important to determine whether the presumptive role of CSR-1 22G-RNAs as the "maintenance of self" system (Shirayama et al., 2012) is linked to their role in regulating endogenous gene expression (likely transcriptional modulation or pre-mRNA processing).

A search for endogenous gene regulation by 21U-RNA-dependent 22G-RNAs identified a fraction of WAGO-22G-RNAs depleted in *prg-1* mutant animals (Bagijn et al., 2012; Lee et al., 2012). Importantly, these 22G-RNAs were distinct from those dependent on 26G-RNAs of the Eri pathway (Lee et al., 2012). Sequences with partial complementarity to existing 21U-RNAs could be identified next to these *prg-1*-dependent 22G loci, identifying them as 21U-RNA-dependent 22G-RNAs (Bagijn et al., 2012; Lee et al., 2012; Figure 1.5). A few genes were found to be negatively regulated by these 22G-RNAs (Bagijn et al., 2012; Lee et al., 2012), although the physiological consequences of this regulation are not clear. Computational prediction of the target sequences antisense to 21U-RNAs, which allowed mismatches, suggested that they are depleted in protein-coding genes (Bagijn et al., 2012), especially CSR-1 targets (Lee et al., 2012). A more recent study, which analyzed spermatogenesis-enriched and oogenesis-enriched 21U-RNAs separately, concluded that the predicted targets of the former were significantly depleted of spermatogenesis genes, while no bias was noted for the predicted targets on the latter class (Shi et al., 2013).

Although the transgene-silencing studies described earlier indicate that 21U-RNAs have the potential to silence foreign DNA elements, it is surprising that so little evidence of this can be found in the *C. elegans* genome. Also, none of the recent studies explained the fertility defects observed in *prg-1* mutant worms and the poor rescue of this phenotype by the catalytically dead PRG-1 protein (Lee et al., 2012), which is very efficient in promoting transgene silencing. Until the mechanistic connection between PRG-1 and its targets clearly explains the progressive sterility phenotype of *prg-1* mutant worms, it is too early to conclude that the role of 21U-RNAs in the biology of *C. elegans* is understood.

5. SYSTEMIC FEATURES OF RNAi

Historically, the ability of dsRNA-induced silencing to spread from cell to cell represented a significant feature of the RNAi phenomenon in *C. elegans* (Fire et al., 1998). The systemic nature of RNAi was also supported by the ability to induce silencing by feeding nematodes dsRNA-expressing bacteria (Timmons & Fire, 1998). However, subsequent studies distinguished "systemic RNAi" from "environmental RNAi" induced by feeding as some mutants were found to be competent in the former and deficient in the latter (reviewed in Whangbo & Hunter, 2008; Zhuang & Hunter, 2012). This section will discuss both types of processes.

5.1. dsRNA import channel SID-1 and the features of mobile RNA species

A screen for systemic RNAi defective (Sid) mutants identified many alleles in the gene *sid-1*, which encodes a conserved protein with multiple transmembrane domains and is required for dsRNA import into cells (Feinberg & Hunter, 2003; Shih, Fitzgerald, Sutherlin, & Hunter, 2009; Shih & Hunter, 2011; Winston, Molodowitch, & Hunter, 2002). The features of the dsRNA import by SID-1 were mostly identified in *Drosophila* S2 cells expressing the *C. elegans* protein (*Drosophila* lacks SID-1 homologs) (Feinberg & Hunter, 2003; Shih et al., 2009; Shih & Hunter, 2011). It was determined that SID-1 is likely to multimerize and to form a dsRNA-gated channel selective for dsRNA (Shih et al., 2009; Shih & Hunter, 2011; Figure 1.8). dsRNA transport through SID-1 occurs by passive diffusion (Feinberg & Hunter, 2003; Shih et al., 2009), and proteins required for the initiation of RNAi prevent the export of imported dsRNA (Shih et al., 2009; Shih & Hunter, 2011). Interestingly, it was shown that transport of dsRNA molecules with single-stranded regions, such as miRNA precursors, is supported by SID-1 (Shih & Hunter, 2011).

Elegant mosaic analyses in *C. elegans* utilizing mutant rescue in specific cells revealed that *sid-1* is not required for the export of systemic silencing RNAs from varying cell types (Jose, Smith, & Hunter, 2009). This study also

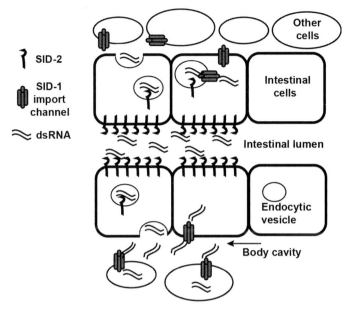

Figure 1.8 Systemic transport of ingested dsRNA. dsRNA is ingested with bacteria and accumulates in the intestinal lumen. SID-2 transmembrane protein is present on the apical side of intestinal cells facing the lumen and is required for the endocytosis of dsRNA. SID-1 is a dsRNA-specific channel composed of several subunits. SID-1 is required for the import of the dsRNA into the cells from endocytic vesicles and from the intercellular space (body cavity). Intestinal cells deficient in SID-1 can support endocytosis-dependent dsRNA transport from the intestinal lumen to body cavity, but SID-1 is required for further import into other cells. *This figure is based on figure 6 in McEwan, Weisman, and Hunter (2012).*

showed that multicopy transgenes readily generate such silencing signals, which spread from cell to cell via SID-1–dependent import (Jose, Smith, & Hunter, 2009). Earlier genetic studies demonstrated that *rde-4* and *rde-1* functions are not required for the generation of the systemic signal, which suggested that long dsRNA could be transported (Tabara et al., 1999). The competence of long dsRNA in systemic transport was further confirmed in mosaic analyses (Jose et al., 2011). This work also revealed the existence of additional mobile RNA species that require *dcr-1* and *rde-4*, but not *rde-1*, for their production, which suggests that double-stranded primary siRNAs, not the single-stranded siRNA products generated after passenger-strand cleavage by RDE-1, are mobile (Jose et al., 2011). Also, it was found that secondary siRNAs generated by RdRP could not be transported (Jose et al., 2011). The nonautonomous role of RDE-4 and the cell-autonomous role of RDE-1 in RNAi were also observed in an earlier study (Blanchard et al., 2011). Surprisingly, the putative nucleotidyltransferase MUT-2 (RDE-3) was required for the generation of a mobile RNA signal; therefore, there is a possibility that primary siRNAs generated after Dicer cleavage can be modified (Jose et al., 2011).

5.2. Additional factors required for systemic and environmental RNAi

5.2.1 SID-2

SID-2 is a transmembrane protein found only in the apical membrane of intestinal cells facing the intestinal lumen (Figure 1.8) and specifically required for the environmental, but not systemic, RNAi response (Winston, Sutherlin, Wright, Feinberg, & Hunter, 2007). Similar proteins exist in other nematode species, such as *C. briggsae* and *C. remanei*, but only *C. elegans* SID-2 is competent in supporting environmental RNAi and can enable this response in *C. briggsae* (Nuez & Felix, 2012; Winston et al., 2007) and *C. remanei* (Nuez & Felix, 2012) when expressed from a transgene. Interestingly, sensitivity to environmental RNAi was observed in several other *Caenorhabditis* species (Nuez & Felix, 2012; Winston et al., 2007), and their phylogenetic analysis is consistent with the convergent evolution of this feature (Nuez & Felix, 2012).

SID-2 is required for the import of environmental dsRNA into intestinal cells, but its function is not sufficient for this import, which also requires SID-1 (Winston et al., 2007). Recent studies conducted in *C. elegans* and *Drosophila* S2 cells suggest a role for SID-2 in the initial uptake of dsRNA from the intestinal lumen by endocytosis and the subsequent requirement of SID-1 for the transport of dsRNA from the endocytic vesicles and/or the body cavity (Figure 1.8; McEwan et al., 2012).

5.2.2 SID-3

SID-3 is a conserved tyrosine kinase related to the Cdc-42-associated kinase (Ack) family and required for the efficient import of dsRNA (Jose, Kim, Leal-Ekman, & Hunter, 2012). It is widely expressed in a variety of *C. elegans* cells, where it shows a punctate pattern of expression. Interestingly, the cell-autonomous RNAi process is not impaired, and even enhanced, in *sid-3* mutant cells (Jose et al., 2012). The kinase domain of SID-3 is required for its function, which suggests that signaling events can modulate the efficiency of dsRNA import into cells (Jose et al., 2012).

5.2.3 SID-5

Mutations in *sid-5* were shown to reduce the efficiency of systemic RNAi (Hinas, Wright, & Hunter, 2012). SID-5 is widely expressed and associates with endosomes; however, unlike *sid-3*, *sid-5* is not required for the import of the silencing signal but appears to affect its export (Hinas et al., 2012). Although the SID-1 channel is required for dsRNA import into cells that execute silencing, a *sid-1*-independent transport of environmental silencing signals across the intestine has been observed (Jose et al., 2009). Such signals, that is, dsRNA, are taken up by SID-2-dependent endocytosis and thought to be released from endocytic vesicles into the body cavity without entry into the intestinal cells (Jose et al., 2009; Figure 1.8). SID-5 was implicated in this *sid-1*-independent transport across the intestine, which is consistent with SID-5's localization to endosomes (Hinas et al., 2012). It is likely that SID-5 generally contributes to systemic RNAi by facilitating the dsRNA export process from a variety of cells.

6. OUTLOOK

This chapter has highlighted the tremendous importance of short RNAs for *C. elegans* biology: short RNAs are used to regulate development, physiology, life span, and stress resistance; they provide an antiviral response and control gene duplications, silence transposons, and modulate expression of endogenous genes by a variety of mechanisms. It is certain that studies of RNAi-related processes in *C. elegans* will continue to thrive and that a comprehensive review of this kind will probably not be possible in the future. It can be expected that a better mechanistic understanding of posttranscriptional and transcriptional gene regulation by short RNAs will be achieved, and that it will be connected in due course to insights about the biological functions of miRNAs, endo-siRNAs, and 21U-RNAs. More often than not, insights from biological systems like *C. elegans* become very relevant to higher organisms and ultimately lead to a better understanding of life.

ACKNOWLEDGMENTS

I am thankful to the members of my lab for many discussions, to Lyuda Kovalchuke for editing the chapter, and to the NIH, the Arnold and Mabel Beckman Foundation, the Leukemia and Lymphoma Society, and the Irma T. Hirschl/Monique Weill-Caulier Trust for funding. Due to the extensive subject matter of this review, references to primary literature were mostly limited to research in *C. elegans*. I apologize to colleagues whose work was not cited.

REFERENCES

Abbott, A. L. (2011). Uncovering new functions for microRNAs in Caenorhabditis elegans. *Current Biology*, *21*(17), R668–R671.

Abbott, A. L., Alvarez-Saavedra, E., Miska, E. A., Lau, N. C., Bartel, D. P., Horvitz, H. R., et al. (2005). The let-7 MicroRNA family members mir-48, mir-84, and mir-241 function together to regulate developmental timing in Caenorhabditis elegans. *Developmental Cell*, *9*(3), 403–414.

Alcazar, R. M., Lin, R., & Fire, A. Z. (2008). Transmission dynamics of heritable silencing induced by double-stranded RNA in Caenorhabditis elegans. *Genetics*, *180*(3), 1275–1288.

Alvarez-Saavedra, E., & Horvitz, H. R. (2010). Many families of C. elegans microRNAs are not essential for development or viability. *Current Biology*, *20*(4), 367–373.

Ambros, V. (2011). MicroRNAs and developmental timing. *Current Opinion in Genetics and Development*, *21*(4), 511–517.

Ambros, V., Lee, R. C., Lavanway, A., Williams, P. T., & Jewell, D. (2003). MicroRNAs and other tiny endogenous RNAs in C. elegans. *Current Biology*, *13*(10), 807–818.

Aoki, K., Moriguchi, H., Yoshioka, T., Okawa, K., & Tabara, H. (2007) In vitro analyses of the production and activity of secondary small interfering RNAs in C. elegans. *EMBO Journal*, *26*(24), 5007–5019.

Ashe, A., Sapetschnig, A., Weick, E. M., Mitchell, J., Bagijn, M. P., Cording, A. C., et al. (2012). piRNAs can trigger a multigenerational epigenetic memory in the germline of C. elegans. *Cell*, *150*(1), 88–99.

Asikainen, S., Storvik, M., Lakso, M., & Wong, G. (2007). Whole genome microarray analysis of C. elegans rrf-3 and eri-1 mutants. *FEBS Letters*, *581*(26), 5050–5054.

Avgousti, D. C., Palani, S., Sherman, Y., & Grishok, A. (2012). CSR-1 RNAi pathway positively regulates histone expression in C. elegans. *EMBO Journal*, *31*(19), 3821–3832.

Bagijn, M. P., Goldstein, L. D., Sapetschnig, A., Weick, E. M., Bouasker, S., Lehrbach, N. J., et al. (2012). Function, targets, and evolution of Caenorhabditis elegans piRNAs. *Science*, *337*(6094), 574–578.

Bartel, D. P. (2009). MicroRNAs: Target recognition and regulatory functions. *Cell*, *136*(2), 215–233.

Batista, P. J., Ruby, J. G., Claycomb, J. M., Chiang, R., Fahlgren, N., Kasschau, K. D., et al. (2008). PRG-1 and 21U-RNAs interact to form the piRNA complex required for fertility in C. elegans. *Molecular Cell*, *31*, 67–78.

Billi, A. C., Alessi, A. F., Khivansara, V., Han, T., Freeberg, M., Mitani, S., et al. (2012). The Caenorhabditis elegans HEN1 ortholog, HENN-1, methylates and stabilizes select subclasses of germline small RNAs. *PLoS Genetics*, *8*(4), e1002617.

Billi, A. C., Freeberg, M. A., Day, A. M., Chun, S. Y., Khivansara, V., & Kim, J. K. (2013). A conserved upstream motif orchestrates autonomous, germline-enriched expression of Caenorhabditis elegans piRNAs. *PLoS Genetics*, *9*(3), e1003392.

Billi, A. C., Freeberg, M. A., & Kim, J. K. (2012). piRNAs and siRNAs collaborate in Caenorhabditis elegans genome defense. *Genome Biology*, *13*(7), 164.

Blanchard, D., Parameswaran, P., Lopez-Molina, J., Gent, J., Saynuk, J. F., & Fire, A. (2011). On the nature of in vivo requirements for rde-4 in RNAi and developmental pathways in C. elegans. *RNA Biology*, *8*(3), 458–467.

Boehm, M., & Slack, F. (2005). A developmental timing microRNA and its target regulate life span in C. elegans. *Science*, *310*(5756), 1954–1957.

Bosher, J. M., Dufourcq, P., Sookhareea, S., & Labouesse, M. (1999). RNA interference can target pre-mRNA: Consequences for gene expression in a Caenorhabditis elegans operon. *Genetics*, *153*(3), 1245–1256.

Bouasker, S., & Simard, M. J. (2012). The slicing activity of miRNA-specific Argonautes is essential for the miRNA pathway in C. elegans. *Nucleic Acids Research*, *40*(20), 10452–10462.

Boulias, K., & Horvitz, H. R. (2012). The C. elegans microRNA mir-71 acts in neurons to promote germline-mediated longevity through regulation of DAF-16/FOXO. *Cell Metabolism, 15*(4), 439–450.

Brenner, J. L., Jasiewicz, K. L., Fahley, A. F., Kemp, B. J., & Abbott, A. L. (2010). Loss of individual microRNAs causes mutant phenotypes in sensitized genetic backgrounds in C. elegans. *Current Biology, 20*(14), 1321–1325.

Brenner, J. L., Kemp, B. J., & Abbott, A. L. (2012). The mir-51 family of microRNAs functions in diverse regulatory pathways in Caenorhabditis elegans. *PLoS One, 7*(5), e37185.

Buckley, B. A., Burkhart, K. B., Gu, S. G., Spracklin, G., Kershner, A., Fritz, H., et al. (2012). A nuclear Argonaute promotes multigenerational epigenetic inheritance and germline immortality. *Nature, 489*(7416), 447–451.

Burkhart, K. B., Guang, S., Buckley, B. A., Wong, L., Bochner, A. F., & Kennedy, S. (2011). A pre-mRNA-associating factor links endogenous siRNAs to chromatin regulation. *PLoS Genetics, 7*(8), e1002249.

Burton, N. O., Burkhart, K. B., & Kennedy, S. (2011). Nuclear RNAi maintains heritable gene silencing in Caenorhabditis elegans. *Proceedings of the National Academy of Sciences of the United States of America, 108*(49), 19683–19688.

Cecere, G., Zheng, G. X., Mansisidor, A. R., Klymko, K. E., & Grishok, A. (2012). Promoters recognized by forkhead proteins exist for individual 21U-RNAs. *Molecular Cell, 47*(5), 734–745.

Ceol, C. J., & Horvitz, H. R. (2004). A new class of C. elegans synMuv genes implicates a Tip60/NuA4-like HAT complex as a negative regulator of Ras signaling. *Developmental Cell, 6*(4), 563–576.

Clark, A. M., Goldstein, L. D., Tevlin, M., Tavare, S., Shaham, S., & Miska, E. A. (2010). The microRNA miR-124 controls gene expression in the sensory nervous system of Caenorhabditis elegans. *Nucleic Acids Research, 38*(11), 3780–3793.

Claycomb, J. M., Batista, P. J., Pang, K. M., Gu, W., Vasale, J. J., van Wolfswinkel, J. C., et al. (2009). The Argonaute CSR-1 and its 22G-RNA cofactors are required for holocentric chromosome segregation. *Cell, 139*(1), 123–134.

Cochella, L., & Hobert, O. (2012). Embryonic priming of a miRNA locus predetermines postmitotic neuronal left/right asymmetry in C. elegans. *Cell, 151*(6), 1229–1242.

Collins, J., Forbes, E., & Anderson, P. (1989). The Tc3 family of transposable genetic elements in Caenorhabditis elegans. *Genetics, 121*(1), 47–55.

Conine, C. C., Batista, P. J., Gu, W., Claycomb, J. M., Chaves, D. A., Shirayama, M., et al. (2010). Argonautes ALG-3 and ALG-4 are required for spermatogenesis-specific 26G-RNAs and thermotolerant sperm in Caenorhabditis elegans. *Proceedings of the National Academy of Sciences of the United States of America, 107*(8), 3588–3593.

Correa, R. L., Steiner, F. A., Berezikov, E., & Ketting, R. F. (2010). MicroRNA-directed siRNA biogenesis in Caenorhabditis elegans. *PLoS Genetics, 6*(4), e1000903.

Cox, D. N., Chao, A., Baker, J., Chang, L., Qiao, D., & Lin, H. (1998). A novel class of evolutionarily conserved genes defined by piwi are essential for stem cell self-renewal. *Genes & Development, 12*(23), 3715–3727.

Cui, M., Chen, J., Myers, T. R., Hwang, B. J., Sternberg, P. W., Greenwald, I., et al. (2006). SynMuv genes redundantly inhibit lin-3/EGF expression to prevent inappropriate vulval induction in C. elegans. *Developmental Cell, 10*(5), 667–672.

Cui, M., Kim, E. B., & Han, M. (2006). Diverse chromatin remodeling genes antagonize the Rb-involved SynMuv pathways in C. elegans. *PLoS Genetics, 2*(5), e74.

Czech, B., & Hannon, G. J. (2011). Small RNA sorting: Matchmaking for Argonautes. *Nature Reviews. Genetics, 12*(1), 19–31.

Das, P. P., Bagijn, M. P., Goldstein, L. D., Woolford, J. R., Lehrbach, N. J., Sapetschnig, A., et al. (2008). Piwi and piRNAs act upstream of an endogenous siRNA pathway to suppress Tc3 transposon mobility in the Caenorhabditis elegans germline. *Molecular Cell*, *31*(1), 79–90.

Davila Lopez, M., & Samuelsson, T. (2008). Early evolution of histone mRNA 3′ end processing. *RNA*, *14*(1), 1–10.

de Lencastre, A., Pincus, Z., Zhou, K., Kato, M., Lee, S. S., & Slack, F. J. (2010). Micro-RNAs both promote and antagonize longevity in C. elegans. *Current Biology*, *20*(24), 2159–2168.

Duchaine, T. F., Wohlschlegel, J. A., Kennedy, S., Bei, Y., Conte, D., Jr., Pang, K., et al. (2006). Functional proteomics reveals the biochemical niche of C. elegans DCR-1 in multiple small-RNA-mediated pathways. *Cell*, *124*(2), 343–354.

Dudley, N. R., Labbe, J. C., & Goldstein, B. (2002). Using RNA interference to identify genes required for RNA interference. *Proceedings of the National Academy of Sciences of the United States of America*, *99*(7), 4191–4196.

Fabian, M. R., Sonenberg, N., & Filipowicz, W. (2010). Regulation of mRNA translation and stability by microRNAs. *Annual Review of Biochemistry*, *79*, 351–379.

Feinberg, E. H., & Hunter, C. P. (2003). Transport of dsRNA into cells by the transmembrane protein SID-1. *Science*, *301*(5639), 1545–1547.

Felix, M. A., Ashe, A., Piffaretti, J., Wu, G., Nuez, I., Belicard, T., et al. (2011). Natural and experimental infection of Caenorhabditis nematodes by novel viruses related to nodaviruses. *PLoS Biology*, *9*(1), e1000586.

Ferguson, E. L., & Horvitz, H. R. (1989). The multivulva phenotype of certain Caenorhabditis elegans mutants results from defects in two functionally redundant pathways. *Genetics*, *123*(1), 109–121.

Fire, A., Xu, S., Montgomery, M. K., Kostas, S. A., Driver, S. E., & Mello, C. C. (1998). Potent and specific genetic interference by double-stranded RNA in Caenorhabditis elegans. *Nature*, *391*(6669), 806–811.

Fischer, S. E. (2010). Small RNA-mediated gene silencing pathways in C. elegans. *The International Journal of Biochemistry & Cell Biology*, *42*(8), 1306–1315.

Fischer, S. E., Butler, M. D., Pan, Q., & Ruvkun, G. (2008). Trans-splicing in C. elegans generates the negative RNAi regulator ERI-6/7. *Nature*, *455*(7212), 491–496.

Fischer, S. E., Montgomery, T. A., Zhang, C., Fahlgren, N., Breen, P. C., Hwang, A., et al. (2011). The ERI-6/7 helicase acts at the first stage of an siRNA amplification pathway that targets recent gene duplications. *PLoS Genetics*, *7*(11), e1002369.

Flamand, M., & Duchaine, T. F. (2012). SnapShot: Endogenous RNAi machinery and mechanisms. *Cell*, *150*(3), 662–662.e2.

Friend, K., Campbell, Z. T., Cooke, A., Kroll-Conner, P., Wickens, M. P., & Kimble, J. (2012). A conserved PUF-Ago-eEF1A complex attenuates translation elongation. *Nature Structural and Molecular Biology*, *19*(2), 176–183.

Gabel, H. W., & Ruvkun, G. (2008). The exonuclease ERI-1 has a conserved dual role in 5.8S rRNA processing and RNAi. *Nature Structural and Molecular Biology*, *15*(5), 531–533.

Gaydos, L. J., Rechtsteiner, A., Egelhofer, T. A., Carroll, C. R., & Strome, S. (2012). Antagonism between MES-4 and Polycomb repressive complex 2 promotes appropriate gene expression in C. elegans germ cells. *Cell Reports*, *2*(5), 1169–1177.

Gent, J. I., Lamm, A. T., Pavelec, D. M., Maniar, J. M., Parameswaran, P., Tao, L., et al. (2010). Distinct phases of siRNA synthesis in an endogenous RNAi pathway in C. elegans soma. *Molecular Cell*, *37*(5), 679–689.

Gent, J. I., Schvarzstein, M., Villeneuve, A. M., Gu, S. G., Jantsch, V., Fire, A. Z., et al. (2009). A Caenorhabditis elegans RNA-directed RNA polymerase in sperm development and endogenous RNA interference. *Genetics*, *183*(4), 1297–1314.

Grishok, A. (2005). RNAi mechanisms in Caenorhabditis elegans. *FEBS Letters*, *579*(26), 5932–5939.

Grishok, A., Hoersch, S., & Sharp, P. A. (2008). RNA interference and retinoblastoma-related genes are required for repression of endogenous siRNA targets in Caenorhabditis elegans. *Proceedings of the National Academy of Sciences of the United States of America, 105*(51), 20386–20391.

Grishok, A., & Mello, C. C. (2002). RNAi (Nematodes: Caenorhabditis elegans). *Advances in Genetics, 46,* 339–360.

Grishok, A., Pasquinelli, A. E., Conte, D., Li, N., Parrish, S., Ha, I., et al. (2001). Genes and mechanisms related to RNA interference regulate expression of the small temporal RNAs that control C. elegans developmental timing. *Cell, 106*(1), 23–34.

Grishok, A., & Sharp, P. A. (2005). Negative regulation of nuclear divisions in Caenorhabditis elegans by retinoblastoma and RNA interference-related genes. *Proceedings of the National Academy of Sciences of the United States of America, 102*(48), 17360–17365.

Grishok, A., Sinskey, J. L., & Sharp, P. A. (2005). Transcriptional silencing of a transgene by RNAi in the soma of C. elegans. *Genes & Development, 19*(6), 683–696.

Grishok, A., Tabara, H., & Mello, C. C. (2000). Genetic requirements for inheritance of RNAi in C. elegans. *Science, 287*(5462), 2494–2497.

Grosshans, H., & Chatterjee, S. (2010). MicroRNAses and the regulated degradation of mature animal miRNAs. *Advances in Experimental Medicine and Biology, 700,* 140–155.

Gu, W., Lee, H. C., Chaves, D., Youngman, E. M., Pazour, G. J., Conte, D., Jr., et al. (2012). CapSeq and CIP-TAP identify Pol II start sites and reveal capped small RNAs as C. elegans piRNA precursors. *Cell, 151*(7), 1488–1500.

Gu, S. G., Pak, J., Barberan-Soler, S., Ali, M., Fire, A., & Zahler, A. M. (2007). Distinct ribonucleoprotein reservoirs for microRNA and siRNA populations in C. elegans. *RNA, 13*(9), 1492–1504.

Gu, S. G., Pak, J., Guang, S., Maniar, J. M., Kennedy, S., & Fire, A. (2012). Amplification of siRNA in Caenorhabditis elegans generates a transgenerational sequence-targeted histone H3 lysine 9 methylation footprint. *Nature Genetics, 44*(2), 157–164.

Gu, W., Shirayama, M., Conte, D., Jr., Vasale, J., Batista, P. J., Claycomb, J. M., et al. (2009). Distinct argonaute-mediated 22G-RNA pathways direct genome surveillance in the C. elegans germline. *Molecular Cell, 36*(2), 231–244.

Guang, S., Bochner, A. F., Burkhart, K. B., Burton, N., Pavelec, D. M., & Kennedy, S. (2010). Small regulatory RNAs inhibit RNA polymerase II during the elongation phase of transcription. *Nature, 465*(7301), 1097–1101.

Guang, S., Bochner, A. F., Pavelec, D. M., Burkhart, K. B., Harding, S., Lachowiec, J., et al. (2008). An Argonaute transports siRNAs from the cytoplasm to the nucleus. *Science, 321*(5888), 537–541.

Gupta, B. P., Hanna-Rose, W., & Sternberg, P. W. (2012). Morphogenesis of the vulva and the vulval-uterine connection. *WormBook,* Nov 30, 1–20.

Habig, J. W., Aruscavage, P. J., & Bass, B. L. (2008). In C. elegans, high levels of dsRNA allow RNAi in the absence of RDE-4. *PLoS One, 3*(12), e4052.

Hall, S. E., Chirn, G. W., Lau, N. C., & Sengupta, P. (2013). RNAi pathways contribute to developmental history-dependent phenotypic plasticity in C. elegans. *RNA, 19*(3), 306–319.

Hammell, C. M. (2008). The microRNA-argonaute complex: A platform for mRNA modulation. *RNA Biology, 5*(3), 123–127.

Hammond, S. M., Boettcher, S., Caudy, A. A., Kobayashi, R., & Hannon, G. J. (2001). Argonaute2, a link between genetic and biochemical analyses of RNAi. *Science, 293*(5532), 1146–1150.

Han, T., Manoharan, A. P., Harkins, T. T., Bouffard, P., Fitzpatrick, C., Chu, D. S., et al. (2009). 26G endo-siRNAs regulate spermatogenic and zygotic gene expression in Caenorhabditis elegans. *Proceedings of the National Academy of Sciences of the United States of America, 106*(44), 18674–18679.

Han, W., Sundaram, P., Kenjale, H., Grantham, J., & Timmons, L. (2008). The Caenorhabditis elegans rsd-2 and rsd-6 genes are required for chromosome functions during exposure to unfavorable environments. *Genetics*, *178*(4), 1875–1893.

Harrison, M. M., Ceol, C. J., Lu, X., & Horvitz, H. R. (2006). Some C. elegans class B synthetic multivulva proteins encode a conserved LIN-35 Rb-containing complex distinct from a NuRD-like complex. *Proceedings of the National Academy of Sciences of the United States of America*, *103*(45), 16782–16787.

Hayes, G. D., Frand, A. R., & Ruvkun, G. (2006). The mir-84 and let-7 paralogous microRNA genes of Caenorhabditis elegans direct the cessation of molting via the conserved nuclear hormone receptors NHR-23 and NHR-25. *Development*, *133*(23), 4631–4641.

Hayes, G. D., Riedel, C. G., & Ruvkun, G. (2011). The Caenorhabditis elegans SOMI-1 zinc finger protein and SWI/SNF promote regulation of development by the mir-84 microRNA. *Genes & Development*, *25*(19), 2079–2092.

Hinas, A., Wright, A. J., & Hunter, C. P. (2012). SID-5 is an endosome-associated protein required for efficient systemic RNAi in C. elegans. *Current Biology*, *22*(20), 1938–1943.

Hobert, O. (2006). Architecture of a microRNA-controlled gene regulatory network that diversifies neuronal cell fates. *Cold Spring Harbor Symposia on Quantitative Biology*, *71*, 181–188.

Hundley, H. A., Krauchuk, A. A., & Bass, B. L. (2008). C. elegans and H. sapiens mRNAs with edited 3′ UTRs are present on polysomes. *RNA*, *14*(10), 2050–2060.

Hutvagner, G., & Simard, M. J. (2008). Argonaute proteins: Key players in RNA silencing. *Nature Reviews. Molecular Cell Biology*, *9*(1), 22–32.

Hutvagner, G., Simard, M. J., Mello, C. C., & Zamore, P. D. (2004). Sequence-specific inhibition of small RNA function. *PLoS Biology*, *2*(4), E98.

Ibanez-Ventoso, C., & Driscoll, M. (2009). MicroRNAs in C. elegans aging: molecular insurance for robustness? *Current Genomics*, *10*(3), 144–153.

Ibanez-Ventoso, C., Yang, M., Guo, S., Robins, H., Padgett, R. W., & Driscoll, M. (2006). Modulated microRNA expression during adult lifespan in Caenorhabditis elegans. *Aging Cell*, *5*(3), 235–246.

Ishizu, H., Siomi, H., & Siomi, M. C. (2012). Biology of PIWI-interacting RNAs: New insights into biogenesis and function inside and outside of germlines. *Genes & Development*, *26*(21), 2361–2373.

Jannot, G., Boisvert, M. E., Banville, I. H., & Simard, M. J. (2008). Two molecular features contribute to the Argonaute specificity for the microRNA and RNAi pathways in C. elegans. *RNA*, *14*(5), 829–835.

Johnson, S. M., Grosshans, H., Shingara, J., Byrom, M., Jarvis, R., Cheng, A., et al. (2005). RAS is regulated by the let-7 microRNA family. *Cell*, *120*(5), 635–647.

Johnston, R. J., & Hobert, O. (2003). A microRNA controlling left/right neuronal asymmetry in Caenorhabditis elegans. *Nature*, *426*(6968), 845–849.

Jose, A. M., Garcia, G. A., & Hunter, C. P. (2011). Two classes of silencing RNAs move between Caenorhabditis elegans tissues. *Nature Structural and Molecular Biology*, *18*(11), 1184–1188.

Jose, A. M., Kim, Y. A., Leal-Ekman, S., & Hunter, C. P. (2012). Conserved tyrosine kinase promotes the import of silencing RNA into Caenorhabditis elegans cells. *Proceedings of the National Academy of Sciences of the United States of America*, *109*(36), 14520–14525.

Jose, A. M., Smith, J. J., & Hunter, C. P. (2009). Export of RNA silencing from C. elegans tissues does not require the RNA channel SID-1. *Proceedings of the National Academy of Sciences of the United States of America*, *106*(7), 2283–2288.

Kai, Z. S., & Pasquinelli, A. E. (2010). MicroRNA assassins: Factors that regulate the disappearance of miRNAs. *Nature Structural and Molecular Biology*, *17*(1), 5–10.

Kamminga, L. M., van Wolfswinkel, J. C., Luteijn, M. J., Kaaij, L. J., Bagijn, M. P., Sapetschnig, A., et al. (2012). Differential impact of the HEN1 homolog HENN-1 on 21U and 26G RNAs in the germline of Caenorhabditis elegans. *PLoS Genetics*, *8*(7), e1002702.

Karp, X., Hammell, M., Ow, M. C., & Ambros, V. (2011). Effect of life history on micro-RNA expression during C. elegans development. *RNA*, *17*(4), 639–651.

Kato, M., Chen, X., Inukai, S., Zhao, H., & Slack, F. J. (2011). Age-associated changes in expression of small, noncoding RNAs, including microRNAs, in C. elegans. *RNA*, *17*(10), 1804–1820.

Kato, M., Paranjape, T., Muller, R. U., Nallur, S., Gillespie, E., Keane, K., et al. (2009). The mir-34 microRNA is required for the DNA damage response in vivo in C. elegans and in vitro in human breast cancer cells. *Oncogene*, *28*(25), 2419–2424.

Kato, M., & Slack, F. J. (2013). Ageing and the small, non-coding RNA world. *Ageing Research Reviews*, *12*(1), 429–435.

Kaufman, E. J., & Miska, E. A. (2010). The microRNAs of Caenorhabditis elegans. *Seminars in Cell & Developmental Biology*, *21*(7), 728–737.

Kemp, B. J., Allman, E., Immerman, L., Mohnen, M., Peters, M. A., Nehrke, K., et al. (2012). miR-786 regulation of a fatty-acid elongase contributes to rhythmic calcium-wave initiation in C. elegans. *Current Biology*, *22*(23), 2213–2220.

Kennedy, S., Wang, D., & Ruvkun, G. (2004). A conserved siRNA-degrading RNase negatively regulates RNA interference in C. elegans. *Nature*, *427*(6975), 645–649.

Ketting, R. F. (2010). MicroRNA biogenesis and function. An overview. *Advances in Experimental Medicine and Biology*, *700*, 1–14.

Ketting, R. F., Haverkamp, T. H., van Luenen, H. G., & Plasterk, R. H. (1999). Mut-7 of C. elegans, required for transposon silencing and RNA interference, is a homolog of Werner syndrome helicase and RNaseD. *Cell*, *99*(2), 133–141.

Ketting, R. F., & Plasterk, R. H. (2000). A genetic link between co-suppression and RNA interference in C. elegans. *Nature*, *404*(6775), 296–298.

Khurana, J. S., & Theurkauf, W. (2010). piRNAs, transposon silencing, and Drosophila germline development. *The Journal of Cell Biology*, *191*(5), 905–913.

Kim, V. N., Han, J., & Siomi, M. C. (2009). Biogenesis of small RNAs in animals. *Nature Reviews. Molecular Cell Biology*, *10*(2), 126–139.

Knight, S. W., & Bass, B. L. (2002). The role of RNA editing by ADARs in RNAi. *Molecular Cell*, *10*(4), 809–817.

Krol, J., Loedige, I., & Filipowicz, W. (2010). The widespread regulation of microRNA biogenesis, function and decay. *Nature Reviews. Genetics*, *11*(9), 597–610.

Kudron, M., Niu, W., Lu, Z., Wang, G., Gerstein, M., Snyder, M., et al. (2013). Tissue-specific direct targets of Caenorhabditis elegans Rb/E2F dictate distinct somatic and germline programs. *Genome Biology*, *14*(1), R5.

Lagos-Quintana, M., Rauhut, R., Lendeckel, W., & Tuschl, T. (2001). Identification of novel genes coding for small expressed RNAs. *Science*, *294*(5543), 853–858.

Lapierre, L. R., & Hansen, M. (2012). Lessons from C. elegans: Signaling pathways for longevity. *Trends in Endocrinology and Metabolism*, *23*(12), 637–644.

Lau, N. C., Lim, L. P., Weinstein, E. G., & Bartel, D. P. (2001). An abundant class of tiny RNAs with probable regulatory roles in Caenorhabditis elegans. *Science*, *294*(5543), 858–862.

Lee, R. C., & Ambros, V. (2001). An extensive class of small RNAs in Caenorhabditis elegans. *Science*, *294*(5543), 862–864.

Lee, R. C., Feinbaum, R. L., & Ambros, V. (1993). The C. elegans heterochronic gene lin-4 encodes small RNAs with antisense complementarity to lin-14. *Cell*, *75*(5), 843–854.

Lee, H. C., Gu, W., Shirayama, M., Youngman, E., Conte, D., Jr., & Mello, C. C. (2012). C. elegans piRNAs mediate the genome-wide surveillance of germline transcripts. *Cell*, *150*(1), 78–87.

Lee, R. C., Hammell, C. M., & Ambros, V. (2006). Interacting endogenous and exogenous RNAi pathways in Caenorhabditis elegans. *RNA, 12*(4), 589–597.

Lehner, B., Calixto, A., Crombie, C., Tischler, J., Fortunato, A., Chalfie, M., et al. (2006). Loss of LIN-35, the Caenorhabditis elegans ortholog of the tumor suppressor p105Rb, results in enhanced RNA interference. *Genome Biology, 7*(1), R4.

Lehrbach, N. J., Castro, C., Murfitt, K. J., Abreu-Goodger, C., Griffin, J. L., & Miska, E. A. (2012). Post-developmental microRNA expression is required for normal physiology, and regulates aging in parallel to insulin/IGF-1 signaling in C. elegans. *RNA, 18*(12), 2220–2235.

Lehrbach, N. J., & Miska, E. A. (2010). Regulation of pre-miRNA processing. *Advances in Experimental Medicine and Biology, 700*, 67–75.

Li, J., & Greenwald, I. (2010). LIN-14 inhibition of LIN-12 contributes to precision and timing of C. elegans vulval fate patterning. *Current Biology, 20*(20), 1875–1879.

Li, M., Jones-Rhoades, M. W., Lau, N. C., Bartel, D. P., & Rougvie, A. E. (2005). Regulatory mutations of mir-48, a C. elegans let-7 family MicroRNA, cause developmental timing defects. *Developmental Cell, 9*(3), 415–422.

Lim, L. P., Lau, N. C., Weinstein, E. G., Abdelhakim, A., Yekta, S., Rhoades, M. W., et al. (2003). The microRNAs of Caenorhabditis elegans. *Genes & Development, 17*(8), 991–1008.

Lu, R., Maduro, M., Li, F., Li, H. W., Broitman-Maduro, G., Li, W. X., et al. (2005). Animal virus replication and RNAi-mediated antiviral silencing in Caenorhabditis elegans. *Nature, 436*(7053), 1040–1043.

Luteijn, M. J., van Bergeijk, P., Kaaij, L. J., Almeida, M. V., Roovers, E. F., Berezikov, E., et al. (2012). Extremely stable Piwi-induced gene silencing in Caenorhabditis elegans. *EMBO Journal, 31*(16), 3422–3430.

Maine, E. M., Hauth, J., Ratliff, T., Vought, V. E., She, X., & Kelly, W. G. (2005). EGO-1, a putative RNA-dependent RNA polymerase, is required for heterochromatin assembly on unpaired dna during C. elegans meiosis. *Current Biology, 15*(21), 1972–1978.

Maniar, J. M., & Fire, A. Z. (2011). EGO-1, a C. elegans RdRP, modulates gene expression via production of mRNA-templated short antisense RNAs. *Current Biology, 21*(6), 449–459.

Mansisidor, A. R., Cecere, G., Hoersch, S., Jensen, M. B., Kawli, T., Kennedy, L. M., et al. (2011). A conserved PHD finger protein and endogenous RNAi modulate insulin signaling in Caenorhabditis elegans. *PLoS Genetics, 7*(9), e1002299.

Margulies, M., Egholm, M., Altman, W. E., Attiya, S., Bader, J. S., Bemben, L. A., et al. (2005). Genome sequencing in microfabricated high-density picolitre reactors. *Nature, 437*(7057), 376–380.

Marzluff, W. F., Wagner, E. J., & Duronio, R. J. (2008). Metabolism and regulation of canonical histone mRNAs: Life without a poly(A) tail. *Nature Reviews. Genetics, 9*(11), 843–854.

Massirer, K. B., Perez, S. G., Mondol, V., & Pasquinelli, A. E. (2012). The miR-35-41 family of microRNAs regulates RNAi sensitivity in Caenorhabditis elegans. *PLoS Genetics, 8*(3), e1002536.

McEwan, D. L., Weisman, A. S., & Hunter, C. P. (2012). Uptake of extracellular double-stranded RNA by SID-2. *Molecular Cell, 47*(5), 746–754.

McManus, M. T., Petersen, C. P., Haines, B. B., Chen, J., & Sharp, P. A. (2002). Gene silencing using micro-RNA designed hairpins. *RNA, 8*(6), 842–850.

McManus, M. T., & Sharp, P. A. (2002). Gene silencing in mammals by small interfering RNAs. *Nature Reviews. Genetics, 3*(10), 737–747.

Miska, E. A., Alvarez-Saavedra, E., Abbott, A. L., Lau, N. C., Hellman, A. B., McGonagle, S. M., et al. (2007). Most Caenorhabditis elegans microRNAs are individually not essential for development or viability. *PLoS Genetics, 3*(12), e215.

Mondol, V., & Pasquinelli, A. E. (2012). Let's make it happen: The role of let-7 microRNA in development. *Current Topics in Developmental Biology, 99,* 1–30.

Montgomery, T. A., Rim, Y. S., Zhang, C., Dowen, R. H., Phillips, C. M., Fischer, S. E., et al. (2012). PIWI associated siRNAs and piRNAs specifically require the Caenorhabditis elegans HEN1 ortholog henn-1. *PLoS Genetics, 8*(4), e1002616.

Nakamura, M., Ando, R., Nakazawa, T., Yudazono, T., Tsutsumi, N., Hatanaka, N., et al. (2007). Dicer-related drh-3 gene functions in germ-line development by maintenance of chromosomal integrity in Caenorhabditis elegans. *Genes to Cells, 12*(9), 997–1010.

Nuez, I., & Felix, M. A. (2012). Evolution of susceptibility to ingested double-stranded RNAs in Caenorhabditis nematodes. *PLoS One, 7*(1), e29811.

Obsil, T., & Obsilova, V. (2011). Structural basis for DNA recognition by FOXO proteins. *Biochimica et Biophysica Acta, 1813*(11), 1946–1953.

Okamura, K., & Lai, E. C. (2008). Endogenous small interfering RNAs in animals. *Nature Reviews. Molecular Cell Biology, 9*(9), 673–678.

Olsson-Carter, K., & Slack, F. J. (2010). A developmental timing switch promotes axon outgrowth independent of known guidance receptors. *PLoS Genetics, 6*(8), e1001054.

Ouellet, J., & Roy, R. (2007). The lin-35/Rb and RNAi pathways cooperate to regulate a key cell cycle transition in C. elegans. *BMC Developmental Biology, 7,* 38.

Pak, J., & Fire, A. (2007). Distinct populations of primary and secondary effectors during RNAi in C. elegans. *Science, 315*(5809), 241–244.

Pak, J., Maniar, J. M., Mello, C. C., & Fire, A. (2012). Protection from feed-forward amplification in an amplified RNAi mechanism. *Cell, 151*(4), 885–899.

Parker, G. S., Eckert, D. M., & Bass, B. L. (2006). RDE-4 preferentially binds long dsRNA and its dimerization is necessary for cleavage of dsRNA to siRNA. *RNA, 12*(5), 807–818.

Parker, G. S., Maity, T. S., & Bass, B. L. (2008). dsRNA binding properties of RDE-4 and TRBP reflect their distinct roles in RNAi. *Journal of Molecular Biology, 384*(4), 967–979.

Parrish, S., & Fire, A. (2001). Distinct roles for RDE-1 and RDE-4 during RNA interference in Caenorhabditis elegans. *RNA, 7*(10), 1397–1402.

Pasquinelli, A. E. (2012). MicroRNAs and their targets: Recognition, regulation and an emerging reciprocal relationship. *Nature Reviews. Genetics, 13*(4), 271–282.

Pasquinelli, A. E., Reinhart, B. J., Slack, F., Martindale, M. Q., Kuroda, M. I., Maller, B., et al. (2000). Conservation of the sequence and temporal expression of let-7 heterochronic regulatory RNA. *Nature, 408*(6808), 86–89.

Pavelec, D. M., Lachowiec, J., Duchaine, T. F., Smith, H. E., & Kennedy, S. (2009). Requirement for the ERI/DICER complex in endogenous RNA interference and sperm development in Caenorhabditis elegans. *Genetics, 183*(4), 1283–1295.

Phillips, C. M., Montgomery, T. A., Breen, P. C., & Ruvkun, G. (2012). MUT-16 promotes formation of perinuclear mutator foci required for RNA silencing in the C. elegans germline. *Genes & Development, 26*(13), 1433–1444.

Rechtsteiner, A., Ercan, S., Takasaki, T., Phippen, T. M., Egelhofer, T. A., Wang, W., et al. (2010). The histone H3K36 methyltransferase MES-4 acts epigenetically to transmit the memory of germline gene expression to progeny. *PLoS Genetics, 6*(9), e1001091.

Reinhart, B. J., Slack, F. J., Basson, M., Pasquinelli, A. E., Bettinger, J. C., Rougvie, A. E., et al. (2000). The 21-nucleotide let-7 RNA regulates developmental timing in Caenorhabditis elegans. *Nature, 403*(6772), 901–906.

Resnick, T. D., McCulloch, K. A., & Rougvie, A. E. (2010). miRNAs give worms the time of their lives: Small RNAs and temporal control in Caenorhabditis elegans. *Developmental Dynamics, 239*(5), 1477–1489.

Rocheleau, C. E., Cullison, K., Huang, K., Bernstein, Y., Spilker, A. C., & Sundaram, M. V. (2008). The Caenorhabditis elegans ekl (enhancer of ksr-1 lethality) genes include putative components of a germline small RNA pathway. *Genetics, 178*(3), 1431–1443.

Rougvie, A. E. (2005). Intrinsic and extrinsic regulators of developmental timing: From miRNAs to nutritional cues. *Development, 132*(17), 3787–3798.

Ruby, J. G., Jan, C., Player, C., Axtell, M. J., Lee, W., Nusbaum, C., et al. (2006). Large-scale sequencing reveals 21U-RNAs and additional microRNAs and endogenous siRNAs in C. elegans. *Cell, 127*(6), 1193–1207.

Ruvkun, G. (2008). The perfect storm of tiny RNAs. *Nature Medicine, 14*(10), 1041–1045.

Sabin, L. R., Delas, M. J., & Hannon, G. J. (2013). Dogma derailed: The many influences of RNA on the genome. *Molecular Cell, 49*(5), 783–794.

Schmitz, C., Wacker, I., & Hutter, H. (2008). The Fat-like cadherin CDH-4 controls axon fasciculation, cell migration and hypodermis and pharynx development in Caenorhabditis elegans. *Developmental Biology, 316*(2), 249–259.

Schott, D. H., Cureton, D. K., Whelan, S. P., & Hunter, C. P. (2005). An antiviral role for the RNA interference machinery in Caenorhabditis elegans. *Proceedings of the National Academy of Sciences of the United States of America, 102*(51), 18420–18424.

Shaw, W. R., Armisen, J., Lehrbach, N. J., & Miska, E. A. (2010). The conserved miR-51 microRNA family is redundantly required for embryonic development and pharynx attachment in Caenorhabditis elegans. *Genetics, 185*(3), 897–905.

She, X., Xu, X., Fedotov, A., Kelly, W. G., & Maine, E. M. (2009). Regulation of heterochromatin assembly on unpaired chromosomes during Caenorhabditis elegans meiosis by components of a small RNA-mediated pathway. *PLoS Genetics, 5*(8), e1000624.

Shen, Y., Wollam, J., Magner, D., Karalay, O., & Antebi, A. (2012). A steroid receptor-microRNA switch regulates life span in response to signals from the gonad. *Science, 338*(6113), 1472–1476.

Shi, Z., Montgomery, T. A., Qi, Y., & Ruvkun, G. (2013). High-throughput sequencing reveals extraordinary fluidity of miRNA, piRNA, and siRNA pathways in nematodes. *Genome Research, 23*(3), 497–508.

Shih, J. D., Fitzgerald, M. C., Sutherlin, M., & Hunter, C. P. (2009). The SID-1 double-stranded RNA transporter is not selective for dsRNA length. *RNA, 15*(3), 384–390.

Shih, J. D., & Hunter, C. P. (2011). SID-1 is a dsRNA-selective dsRNA-gated channel. *RNA, 17*(6), 1057–1065.

Shirayama, M., Seth, M., Lee, H. C., Gu, W., Ishidate, T., Conte, D., Jr., et al. (2012). piRNAs initiate an epigenetic memory of nonself RNA in the C. elegans germline. *Cell, 150*(1), 65–77.

Sijen, T., Fleenor, J., Simmer, F., Thijssen, K. L., Parrish, S., Timmons, L., et al. (2001). On the role of RNA amplification in dsRNA-triggered gene silencing. *Cell, 107*(4), 465–476.

Sijen, T., & Plasterk, R. H. (2003). Transposon silencing in the Caenorhabditis elegans germ line by natural RNAi. *Nature, 426*(6964), 310–314.

Sijen, T., Steiner, F. A., Thijssen, K. L., & Plasterk, R. H. (2007). Secondary siRNAs result from unprimed RNA synthesis and form a distinct class. *Science, 315*(5809), 244–247.

Simmer, F., Tijsterman, M., Parrish, S., Koushika, S. P., Nonet, M. L., Fire, A., et al. (2002). Loss of the putative RNA-directed RNA polymerase RRF-3 makes C. elegans hypersensitive to RNAi. *Current Biology, 12*(15), 1317–1319.

Simon, D. J., Madison, J. M., Conery, A. L., Thompson-Peer, K. L., Soskis, M., Ruvkun, G. B., et al. (2008). The microRNA miR-1 regulates a MEF-2-dependent retrograde signal at neuromuscular junctions. *Cell, 133*(5), 903–915.

Smardon, A., Spoerke, J. M., Stacey, S. C., Klein, M. E., Mackin, N., & Maine, E. M. (2000). EGO-1 is related to RNA-directed RNA polymerase and functions in germ-line development and RNA interference in C. elegans. *Current Biology, 10*(4), 169–178.

Smith-Vikos, T., & Slack, F. J. (2012). MicroRNAs and their roles in aging. *Journal of Cell Science, 125*(Pt. 1), 7–17.

Sokol, N. S. (2012). Small temporal RNAs in animal development. *Current Opinion in Genetics and Development, 22*(4), 368–373.

Sokol, N. S., & Ambros, V. (2005). Mesodermally expressed Drosophila microRNA-1 is regulated by Twist and is required in muscles during larval growth. *Genes & Development, 19*(19), 2343–2354.

Spike, C. A., Bader, J., Reinke, V., & Strome, S. (2008). DEPS-1 promotes P-granule assembly and RNA interference in C. elegans germ cells. *Development, 135*(5), 983–993.

Steiner, F. A., Hoogstrate, S. W., Okihara, K. L., Thijssen, K. L., Ketting, R. F., Plasterk, R. H., et al. (2007). Structural features of small RNA precursors determine Argonaute loading in Caenorhabditis elegans. *Nature Structural and Molecular Biology, 14*(10), 927–933.

Steiner, F. A., Okihara, K. L., Hoogstrate, S. W., Sijen, T., & Ketting, R. F. (2009). RDE-1 slicer activity is required only for passenger-strand cleavage during RNAi in Caenorhabditis elegans. *Nature Structural and Molecular Biology, 16*(2), 207–211.

Sundaram, P., Echalier, B., Han, W., Hull, D., & Timmons, L. (2006). ATP-binding cassette transporters are required for efficient RNA interference in Caenorhabditis elegans. *Molecular Biology of the Cell, 17*(8), 3678–3688.

Sundaram, P., Han, W., Cohen, N., Echalier, B., Albin, J., & Timmons, L. (2008). Caenorhabditis elegans ABCRNAi transporters interact genetically with rde-2 and mut-7. *Genetics, 178*(2), 801–814.

Tabach, Y., Billi, A. C., Hayes, G. D., Newman, M. A., Zuk, O., Gabel, H., et al. (2013). Identification of small RNA pathway genes using patterns of phylogenetic conservation and divergence. *Nature, 493*(7434), 694–698.

Tabara, H., Sarkissian, M., Kelly, W. G., Fleenor, J., Grishok, A., Timmons, L., et al. (1999). The rde-1 gene, RNA interference, and transposon silencing in C. elegans. *Cell, 99*(2), 123–132.

Tabara, H., Yigit, E., Siomi, H., & Mello, C. C. (2002). The dsRNA binding protein RDE-4 interacts with RDE-1, DCR-1, and a DExH-box helicase to direct RNAi in C. elegans. *Cell, 109*(7), 861–871.

Thivierge, C., Makil, N., Flamand, M., Vasale, J. J., Mello, C. C., Wohlschlegel, J., et al. (2012). Tudor domain ERI-5 tethers an RNA-dependent RNA polymerase to DCR-1 to potentiate endo-RNAi. *Nature Structural and Molecular Biology, 19*(1), 90–97.

Tijsterman, M., Ketting, R. F., Okihara, K. L., Sijen, T., & Plasterk, R. H. (2002). RNA helicase MUT-14-dependent gene silencing triggered in C. elegans by short antisense RNAs. *Science, 295*(5555), 694–697.

Tijsterman, M., May, R. C., Simmer, F., Okihara, K. L., & Plasterk, R. H. (2004). Genes required for systemic RNA interference in Caenorhabditis elegans. *Current Biology, 14*(2), 111–116.

Tijsterman, M., Okihara, K. L., Thijssen, K., & Plasterk, R. H. (2002). PPW-1, a PAZ/PIWI protein required for efficient germline RNAi, is defective in a natural isolate of C. elegans. *Current Biology, 12*(17), 1535–1540.

Timmons, L., & Fire, A. (1998). Specific interference by ingested dsRNA. *Nature, 395*(6705), 854.

Tonkin, L. A., Saccomanno, L., Morse, D. P., Brodigan, T., Krause, M., & Bass, B. L. (2002). RNA editing by ADARs is important for normal behavior in Caenorhabditis elegans. *EMBO Journal, 21*(22), 6025–6035.

Tops, B. B., Tabara, H., Sijen, T., Simmer, F., Mello, C. C., Plasterk, R. H., et al. (2005). RDE-2 interacts with MUT-7 to mediate RNA interference in Caenorhabditis elegans. *Nucleic Acids Research, 33*(1), 347–355.

Turner, M. J., & Slack, F. J. (2009). Transcriptional control of microRNA expression in C. elegans: Promoting better understanding. *RNA Biology, 6*(1), 49–53.

Unhavaithaya, Y., Shin, T. H., Miliaras, N., Lee, J., Oyama, T., & Mello, C. C. (2002). MEP-1 and a homolog of the NURD complex component Mi-2 act together to maintain germline-soma distinctions in C. elegans. *Cell, 111*(7), 991–1002.

Updike, D. L., & Strome, S. (2009). A genomewide RNAi screen for genes that affect the stability, distribution and function of P granules in Caenorhabditis elegans. *Genetics*, *183*(4), 1397–1419.

van Wolfswinkel, J. C., Claycomb, J. M., Batista, P. J., Mello, C. C., Berezikov, E., & Ketting, R. F. (2009). CDE-1 affects chromosome segregation through uridylation of CSR-1-bound siRNAs. *Cell*, *139*(1), 135–148.

Vasale, J. J., Gu, W., Thivierge, C., Batista, P. J., Claycomb, J. M., Youngman, E. M., et al. (2010). Sequential rounds of RNA-dependent RNA transcription drive endogenous small-RNA biogenesis in the ERGO-1/Argonaute pathway. *Proceedings of the National Academy of Sciences of the United States of America*, *107*(8), 3582–3587.

Vasquez-Rifo, A., Jannot, G., Armisen, J., Labouesse, M., Bukhari, S. I., Rondeau, E. L., et al. (2012). Developmental characterization of the microRNA-specific C. elegans Argonautes alg-1 and alg-2. *PLoS One*, *7*(3), e33750.

Vastenhouw, N. L., Brunschwig, K., Okihara, K. L., Muller, F., Tijsterman, M., & Plasterk, R. H. (2006). Gene expression: Long-term gene silencing by RNAi. *Nature*, *442*(7105), 882.

Vastenhouw, N. L., Fischer, S. E., Robert, V. J., Thijssen, K. L., Fraser, A. G., Kamath, R. S., et al. (2003). A genome-wide screen identifies 27 genes involved in transposon silencing in C. elegans. *Current Biology*, *13*(15), 1311–1316.

Vella, M. C., & Slack, F. J. (2005). C. elegans microRNAs. *WormBook*, Sep 21, 1–9.

Vought, V. E., Ohmachi, M., Lee, M. H., & Maine, E. M. (2005). EGO-1, a putative RNA-directed RNA polymerase, promotes germline proliferation in parallel with GLP-1/notch signaling and regulates the spatial organization of nuclear pore complexes and germline P granules in Caenorhabditis elegans. *Genetics*, *170*(3), 1121–1132.

Wang, D., Kennedy, S., Conte, D., Jr., Kim, J. K., Gabel, H. W., Kamath, R. S., et al. (2005). Somatic misexpression of germline P granules and enhanced RNA interference in retinoblastoma pathway mutants. *Nature*, *436*(7050), 593–597.

Wang, G., & Reinke, V. (2008). A C. elegans Piwi, PRG-1, regulates 21U-RNAs during spermatogenesis. *Current Biology*, *18*(12), 861–867.

Warf, M. B., Shepherd, B. A., Johnson, W. E., & Bass, B. L. (2012). Effects of ADARs on small RNA processing pathways in C. elegans. *Genome Research*, *22*(8), 1488–1498.

Welker, N. C., Habig, J. W., & Bass, B. L. (2007). Genes misregulated in C. elegans deficient in Dicer, RDE-4, or RDE-1 are enriched for innate immunity genes. *RNA*, *13*(7), 1090–1102.

Welker, N. C., Maity, T. S., Ye, X., Aruscavage, P. J., Krauchuk, A. A., Liu, Q., et al. (2011). Dicer's helicase domain discriminates dsRNA termini to promote an altered reaction mode. *Molecular Cell*, *41*(5), 589–599.

Welker, N. C., Pavelec, D. M., Nix, D. A., Duchaine, T. F., Kennedy, S., & Bass, B. L. (2010). Dicer's helicase domain is required for accumulation of some, but not all, C. elegans endogenous siRNAs. *RNA*, *16*(5), 893–903.

Whangbo, J. S., & Hunter, C. P. (2008). Environmental RNA interference. *Trends in Genetics*, *24*(6), 297–305.

Wightman, B., Ha, I., & Ruvkun, G. (1993). Posttranscriptional regulation of the heterochronic gene lin-14 by lin-4 mediates temporal pattern formation in C. elegans. *Cell*, *75*(5), 855–862.

Wilkins, C., Dishongh, R., Moore, S. C., Whitt, M. A., Chow, M., & Machaca, K. (2005). RNA interference is an antiviral defence mechanism in Caenorhabditis elegans. *Nature*, *436*(7053), 1044–1047.

Winston, W. M., Molodowitch, C., & Hunter, C. P. (2002). Systemic RNAi in C. elegans requires the putative transmembrane protein SID-1. *Science*, *295*(5564), 2456–2459.

Winston, W. M., Sutherlin, M., Wright, A. J., Feinberg, E. H., & Hunter, C. P. (2007). Caenorhabditis elegans SID-2 is required for environmental RNA interference.

Proceedings of the National Academy of Sciences of the United States of America, 104(25), 10565–10570.

Wu, D., Lamm, A. T., & Fire, A. Z. (2011). Competition between ADAR and RNAi pathways for an extensive class of RNA targets. *Nature Structural and Molecular Biology, 18*(10), 1094–1101.

Wu, X., Shi, Z., Cui, M., Han, M., & Ruvkun, G. (2012). Repression of germline RNAi pathways in somatic cells by retinoblastoma pathway chromatin complexes. *PLoS Genetics, 8*(3), e1002542.

Wu, E., Thivierge, C., Flamand, M., Mathonnet, G., Vashisht, A. A., Wohlschlegel, J., et al. (2010). Pervasive and cooperative deadenylation of 3'UTRs by embryonic microRNA families. *Molecular Cell, 40*(4), 558–570.

Yang, J., Chen, D., He, Y., Melendez, A., Feng, Z., Hong, Q., et al. (2013). MiR-34 modulates Caenorhabditis elegans lifespan via repressing the autophagy gene atg9. *Age (Dordrecht, Netherlands), 35*(1), 11–22.

Yang, H., Zhang, Y., Vallandingham, J., Li, H., Florens, L., & Mak, H. Y. (2012). The RDE-10/RDE-11 complex triggers RNAi-induced mRNA degradation by association with target mRNA in C. elegans. *Genes & Development, 26*(8), 846–856.

Yigit, E., Batista, P. J., Bei, Y., Pang, K. M., Chen, C. C., Tolia, N. H., et al. (2006). Analysis of the C. elegans Argonaute family reveals that distinct Argonautes act sequentially during RNAi. *Cell, 127*(4), 747–757.

Yoo, A. S., & Greenwald, I. (2005). LIN-12/Notch activation leads to microRNA-mediated down-regulation of Vav in C. elegans. *Science, 310*(5752), 1330–1333.

Zhang, C., Montgomery, T. A., Fischer, S. E., Garcia, S. M., Riedel, C. G., Fahlgren, N., et al. (2012). The Caenorhabditis elegans RDE-10/RDE-11 complex regulates RNAi by promoting secondary siRNA amplification. *Current Biology, 22*(10), 881–890.

Zhang, C., Montgomery, T. A., Gabel, H. W., Fischer, S. E., Phillips, C. M., Fahlgren, N., et al. (2011). mut-16 and other mutator class genes modulate 22G and 26G siRNA pathways in Caenorhabditis elegans. *Proceedings of the National Academy of Sciences of the United States of America, 108*(4), 1201–1208.

Zhao, Z., Boyle, T. J., Liu, Z., Murray, J. I., Wood, W. B., & Waterston, R. H. (2010). A negative regulatory loop between microRNA and Hox gene controls posterior identities in Caenorhabditis elegans. *PLoS Genetics, 6*(9), e1001089.

Zhao, Y., Ransom, J. F., Li, A., Vedantham, V., von Drehle, M., Muth, A. N., et al. (2007). Dysregulation of cardiogenesis, cardiac conduction, and cell cycle in mice lacking miRNA-1-2. *Cell, 129*(2), 303–317.

Zhuang, J. J., & Hunter, C. P. (2012). RNA interference in Caenorhabditis elegans: Uptake, mechanism, and regulation. *Parasitology, 139*(5), 560–573.

Zisoulis, D. G., Kai, Z. S., Chang, R. K., & Pasquinelli, A. E. (2012). Autoregulation of microRNA biogenesis by let-7 and Argonaute. *Nature, 486*(7404), 541–544.

CHAPTER TWO

Genetics of Immune Recognition and Response in *Drosophila* host defense

Petros Ligoxygakis[1]

Laboratory of Genes and Development, Department of Biochemistry, University of Oxford, Oxford, United Kingdom
[1]Corresponding author: e-mail address: petros.igoxygakis@bioch.ox.ac.uk

Contents

1. Introduction 72
  1.1 Overview of *Drosophila* immune responses 73
2. Toll Signaling in *Drosophila* 75
  2.1 Activation 75
  2.2 Intracellular signaling 81
  2.3 Host–pathogen interaction upstream of Toll 82
3. The Immune Deficiency Pathway 83
  3.1 Recognition and intracellular signaling 83
  3.2 Gut immunity and IMD signaling 85
  3.3 Host–pathogen interaction 87
4. Immunity as a Sensor of Metabolism 87
5. Concluding Remarks—Future Directions 88
Acknowledgments 89
References 89

Abstract

Due to the evolutionary conservation of innate immune mechanisms, *Drosophila* has been extensively used as a model for the dissection in genetic terms of innate host immunity to infection. Genetic screening in fruit flies has set the stage for the pathways and systems required for responding to immune challenge and the dynamics of the progression of bacterial and fungal infection. In addition, fruit flies have been used as infection models to dissect host–pathogen interactions from both sides of this equation. This chapter describes our current understanding of the genetics of the fruit fly immune response and summarizes the most important findings in this area during the past decade.

Advances in Genetics, Volume 83
ISSN 0065-2660
http://dx.doi.org/10.1016/B978-0-12-407675-4.00002-X

1. INTRODUCTION

To explain the paradox for the requirement of microbial constituents in addition to antigen in inducing adaptive immunity, the late great Charles Janeway formulated the pattern recognition hypothesis (Janeway, 1989). This hypothesis had several implications for microbial recognition. Micro-organisms can evolve much faster than their hosts, and therefore the latter (the hypothesis went) would have to recognize microbial molecules that remained unchanged during evolution. It would be the recognition of these molecules by germ-line-coded receptors of the host that would activate adaptive immunity. In their turn, adaptive receptors would have evolved to overcome the constraints of their innate counterparts and target an infinite diversity of antigens so that every pathogen would be specifically countered. In other words, innate immunity would be a recognition system with a broad remit able to induce adaptive responses that would in turn provide specificity.

In 1996, Jules Hoffman and coworkers published a pivotal paper demonstrating that the Toll receptor pathway in *Drosophila* was required for antifungal defense (Lemaitre, Nicolas, Michaut, Reichhart, & Hoffmann, 1996). It was known from studies in early *Drosophila* embryogenesis that Toll signaling was important for dorsal–ventral polarity and culminated in the translocation to the nucleus of a transcription factor homologous to mammalian NF-κB (Steward, 1987). The Lemaitre paper along with the notion that Toll was controlling a conserved downstream cascade stimulated Janeway's group to look for orthologous receptors. Indeed, they found a human homologue of *Drosophila* Toll now known as Toll-like receptor 4 (TLR4), which was signaling through NF-κB to activate proinflammatory cytokines and the costimulatory molecule B7 in the monocytic cell line THP-I (Medzhitov, Preston-Hurlburt, & Janeway, 1997). A year later, there was *in vivo* evidence. The positional cloning of the gene responsible for the *lps* mutation in mice revealed that TLR4 was accountable for the phenotype, namely, mice that were not sensitive to lipopolysaccharides (LPS) but severely susceptible to some Gram-negative bacteria (Poltorak et al., 1998). Finally, mice with a targeted deletion of the TLR4 gene were unresponsive to LPS (Hoshino et al., 1999). Together, these studies demonstrated the essential role for TLR4 in recognition of LPS in Gram-negative bacteria.

We now know that there is a family of 9 Toll receptors in *Drosophila* as well as 12 TLRs in mice and 10 in humans, a fact that underlines the importance of innate immune recognition placing it under an intense evolutionary light (Janeway & Medzhitov, 2002; Medzhitov, 2009). The work of Jean-Laurent Casanova, Laurent Abel, and coworkers in identifying and analyzing the effects of naturally occurring mutations in TLR pathways in humans has closed the circle and linked these receptors to human innate defenses (Casanova, Abel, & Quintana-Murci, 2011). Mutations in TLRs or components of their pathways include NEMO in TLR3-mediated defense against herpes simplex virus (HSV; Audry et al., 2011), Myd88 (see below) in defenses against pyogenic bacteria (Von Bernuth et al., 2008), IRAK-4 in antibacterial responses (Ku et al., 2007), and TLR3 itself against HSV (Zhang et al., 2007). However, studies in humans revealed an interesting aspect compared to data from animal models. Loss of function of TLRs or components of their downstream pathways had a narrower phenotype than the one expected from *Drosophila* or mouse knockouts. This made the case for redundancy displayed by cell-surface TLRs in humans (Bousfiha et al., 2010). Nevertheless, genetic and epidemiological association studies found that certain variations in TLRs in human populations present increased risks for leprosy (TLR1), pulmonary tuberculosis (TL2), meningococcal disease (TLR4), and accelerated progression of HIV-1 (TLR9) (Casanova et al., 2011).

1.1. Overview of *Drosophila* immune responses

Drosophila lacks a closed circulatory system, and following systemic infection pathogens will be in contact with all organs inside the body cavity. Conversely, mediators and effectors of immunity to infection will be disseminated immediately. These include cellular components (blood cells) as well as humoral components involved in melanization (circulating as inactive zymogens) and coagulation cascades and antimicrobial peptides (AMPs). AMPs are produced in the fat body, an analogue of the mammalian liver (Rizki, 1978); AMP activation and specificity have been connected to types of invading pathogens (Irving et al., 2001; Lemaitre, Reichhart, & Hoffmann, 1997), while biochemical characteristics of AMPs have been extensively studied and reviewed (Hetru, Troxler, & Hoffmann, 2003).

Drosophila blood cells (or hemocytes) are considered the insect equivalent to vertebrate blood cells. Recent studies (reviewed in Crozatier & Vincent 2011) along with a classic paper by Hartenstein and colleagues

(Tepass, Fessler, Aziz, & Hartenstein, 1994) have delineated the ontogeny of these cells from embryonic development (plasmatocytes and crystal cells) to larval stages, where they persist and form circulating and sessile subpopulations, and then through metamorphosis to adults. Following the first phase of hematopoiesis in embryos, there is a second phase in larvae directed by a specialized compartmentalized organ situated in the dorsal aorta, called the lymph gland. The size of the lymph gland is regulated by bone morphogenetic protein signaling as well as TOR/insulin activity (Benmimoun, Polesello, Waltzer, & Haenlin, 2012; Pennetier et al., 2012). The lymph gland contains progenitors (prohemocytes) for three types of functional hemocytes including the plasmatocytes, which are monocyte-like cells involved in phagocytosis of apoptotic bodies and pathogens, and crystal cells, which are required for melanization. These two hemocyte types are released in the hemolymph upon dispersal of the lymph gland at the onset of the larva-to-pupa transition. The hematopoietic organ also gives rise to a third type of hemocyte, the lamellocyte, devoted to encapsulation of foreign bodies that are too large to be phagocytosed. Lamellocytes do not differentiate under normal developmental conditions but only in response to specific immune challenges such as wasp parasitism or stress conditions mediated by an increase in ROS. Mutant backgrounds with increased hemocyte proliferation lead to the formation of melanotic "tumors" that result from encapsulation of larval tissue by lamellocytes. In this context, large-scale screens to identify melanotic-tumor-suppressor genes have been published uncovering new genes and gene networks controlling hemocyte homeostasis (Avet-Rochex et al., 2010; Tokusumi, Tokusumi, Shoue, & Schulz, 2012).

Melanization (or the deposition of melanin) is an acute and highly localized defense triggered by wounding and the presence of invading microbes, eukaryotic parasites, or wasp eggs. The biochemistry of the process has been studied extensively in other insects and reviewed elsewhere (Jiang, Vilcinskas, & Kanost, 2010). Central to this reaction is the enzyme Phenoloxidase (an oxidoreductase), which circulates in the blood as an inactive proenzyme and upon activation by an upstream serine protease cascade catalyzes the conversion of phenols to quinones. Quinones may be directly toxic to bacteria, fungi, and eukaryotic parasites and can also polymerize nonenzymatically to form melanotic capsules that surround parasites (Nappi, Poirié, & Carton, 2009). Removal of a serine protease inhibitor (Serpin-27A) that blocks melanization in the absence of immune challenge provokes spontaneous melanization and increased lethality during development

(Ligoxygakis, Pelte, Hoffmann, & Reichhart, 2002a; Ligoxygakis, Pelte, Ji, et al., 2002b). Melanization is coupled with an additional humoral proteolytic cascade, which leads to blood (hemolymph) clotting with the evolutionary conserved protein transglutaminase (TG) having a central role in the entrapment of microbes. TG provides the "bridge" between hemolymph clot and bacterial surfaces although whether TG binds directly to bacteria or connects to a recognition receptor is not yet known (Wang et al., 2010).

The timeline of induction and coordination of blood cell and AMP-dependent responses is very interesting. Following systemic infection, coagulation and melanization are the immediate and acute responses to localize and/or encapsulate infection and clot the wound. Phagocytosis is able to clear more than 95% of bacteria in the first half an hour (Shia et al., 2009). Interestingly, both in *Drosophila* as well as in the beetle *Tenebrio molitor*, AMP levels start to increase only after the vast majority of bacteria have been eliminated (Haine, Moret, Siva-Jothy, & Rolff, 2008; Shia et al., 2009). These results imply that induced antimicrobial compounds function primarily to protect the insect against the bacteria that persist within their body, rather than to clear microbial infections (Haine et al., 2008).

There is no evidence to suggest that insects have an adaptive system like that of mammals: specific antibodies are not produced; no sign of somatic gene rearrangement or a system similar to MHC antigen presentation has been found. There have been studies, however, that implicate the gene encoding Down syndrome cell adhesion molecule (Dscam) contributing to an expanded insect immune response diversity and specificity. This would be achieved through alternative splicing of three immunoglobulin (Ig) domain-encoding exon cassettes that are each alternatively spliced to contribute a single Ig exon in individual mature messenger RNAs (Dong, Cirimotich, Pike, Chandra, & Dimopoulos, 2012; Watson et al., 2005).

2. TOLL SIGNALING IN *DROSOPHILA*

2.1. Activation

2.1.1 The Necrotic–Persephone axis

From the fruit fly perspective, there is an important difference between Toll signaling in *Drosophila* and TLR signaling in mammals. In contrast to the latter where the interaction with microbial molecules is at the level of the receptor, *Drosophila* Toll is activated by an endogenous ligand, namely, Spätzle (Spz) (Weber et al., 2003). Spz is a cytokine-like molecule homologous to the Nerve Growth Factor (DeLotto & DeLotto, 1998). This finding, which was first substantiated in the 1996 Lemaitre paper, meant that recognition of microbial components would happen upstream of receptor activation. In early embryonic development, Spz is proteolytically cleaved and thus activated by a serine protease cascade (LeMosy, Hong, & Hashimoto, 1999). These proteases were found not to be involved in immunity (Lemaitre et al., 1996). So the question how Spz activated Toll during host defense remained. The first evidence for the possible involvement of a proteolytic cascade in the activation of Toll came with the isolation and subsequent characterization of a mutant that knocked out the serine protease inhibitor Necrotic (Nec). The phenotype was death in early adulthood (in the strongest allelic combinations 3–5 days posteclosion), constitutive expression of the AMP gene *drosomycin* (a target of Toll signaling), and spontaneous melanization in the absence of any immune challenge (Levashina et al., 1999). The nature of the protein coded by the mutant gene as well as the Nec phenotype pointed to the existence of a serine protease (or proteases) different from the ones involved in the embryo, which in the absence of Nec would constitutively activate Toll. The latter result also showed how an overreacting immune response was lethal to the host. It also set the stage for a forward genetic screen harnessing the power of *Drosophila* as a model system. If the protease in question were mutated in a Nec background, then (in the simplest possible scenario) all aspects of the Nec phenotype (including death) would be suppressed. Indeed, through chemical mutagenesis, a suppressor was identified, blocking all aspects of the Nec phenotype. Flies carrying the mutation in a Nec homozygous background were as viable as wild-type flies but were very susceptible to fungal infection, failing to activate *drosomycin*. The mutation responsible was in a gene coding for a serine protease that was subsequently named Persephone (Psh) (Ligoxygakis, Pelte, Hoffmann, et al., 2002; Ligoxygakis, Pelte, Ji, et al., 2002). Later, it was shown that Psh was activated by fungal proteases (Gottar et al., 2006). Finally, it was found that Psh was also induced by bacterial proteases and thus acting as a sensor of "danger" for both fungal and bacterial infection (El Chamy, Leclerc, Caldelari, & Reichhart, 2008).

2.1.2 PGRPs and GNBPs

Genetic screens identified additional components upstream of Toll induction acting in parallel to Psh. A mutation in a gene coding for peptidoglycan recognition protein-SA (PGRP-SA) rendered flies susceptible to Gram-positive bacterial infection and unable to activate *drosomycin* (Michel, Reichhart, Hoffmann, & Royet, 2001). This mutant was named

Semmelweis (in honor of the famous Hungarian physician who discovered the cause of puerperal fever). PGRPs belong to a family of proteins divided into two classes: those with short transcripts and 5'-untranslated regions, namely, PGRP-SA, SB1, SB2, SC1A, SC1B, SC2, and SD, and those with long transcripts and long 5'-untranslated regions including PGRP-LA, LB, LC, LD, and LE. The predicted structures indicate that the first group encodes extracellular proteins and the second group, intracellular and membrane-spanning proteins (Werner et al., 2000). Functional data confirmed this prediction and extended both our knowledge and our conceptual grasp of how PGRPs act in the insect immune system (see below). In the context of Toll activation in addition to PGRP-SA, PGRP-SD has been genetically implicated in the reception of some Gram-positive bacteria (Bischoff et al., 2004). Both are predicted to be circulating in the hemolymph (Michel et al., 2001; Wang et al., 2008). An additional family of proteins that have been implicated in pathogen sensing upstream of Toll is the glucan-binding proteins (GNBPs) initially characterized in large-sized lepidopterans (Ma, 2000; Ochiai, 2000). There are three canonical GNBPs in *Drosophila*, namely, GNBPs 1–3. A mutation named *Osiris* (*osi*), corresponding to the insertion of a modified piggyBac transposon in the coding region of GNBP1, rendered flies unable to activate the Toll pathway following infection by *Micrococcus luteus* or *Enterococcus faecalis* and being very sensitive to immune challenge by the latter (Gobert et al., 2003). Direct physical interaction between GNBP1 and PGRP-SA was observed in native protein gels from whole fly extracts (Gobert et al., 2003) but has since measured with biophysical methods using purified proteins in the low micromolar range and found to be significantly enhanced in the presence of muropeptides (Wang et al., 2006).

P-element-mediated partial deletion of the gene coding GNBP3 resulted in inability to activate Toll signaling through fungal cell wall components (using heat-killed *Candida albicans*) and susceptibility to fungal infection (Gottar et al., 2006). Genetically, GNBP3 was *not* found to be upstream of Psh (Gottar et al., 2006). Interestingly, however, when *psh* and *gnbp3* were combined, flies died earlier than the single mutants from fungal infection pointing to the existence of two parallel pathways, one activated by proteases (and mediated by Psh) and the other by fungal cell wall components (mediated by GNBP3) (Gottar et al., 2006). Indeed, a fungal protease from *Metarhizium anisopliae* expressed in flies was able to activate Toll in a Psh-dependent manner (Gottar et al., 2006), whereas GNBP3 bound long chains of β-1,3-glucans in the fungal cell wall of *C. albicans* and *Aspergillus fumigatus* (Mishima et al., 2009).

2.1.3 Mechanism of sensing bacterial and fungal cell wall components

The emerging picture for pathogen recognition upstream of Toll, therefore, is one that has three strands (Figure 2.1). First, bacterial or fungal proteases activate directly or indirectly Psh feeding in the proteolytic cascade leading to Spz cleavage (see below). Second, peptidoglycan (PG) from Gram-positive bacteria is recognized by PGRP-SA, GNBP1, and PGRP-SD (see below), and third, β-glucan from fungal spores is sensed by GNBP3. The mechanism by which pathogen proteases activate Psh and whether this activation is direct or involves another protease upstream of Psh remain unknown. Nevertheless, some progress has been made in exploring the possible ways that sensing may be achieved in the case of cell wall components of bacteria and fungi. This is a pressing issue since it will explain how a small number of germ line-coded receptors can sense the vast amount of variability in the microbial cell wall, especially those of Gram-positive bacteria that (due to their variability) have defied categorization by the pattern recognition theory.

It has become apparent that PGRP-SA is able to bind whole bacteria (Atilano, Yates, Glittenberg, Filipe, & Ligoxygakis, 2011) as well as cell wall and PG fragments (Wang et al., 2006). Which one happens *in vivo* is a big question (and one actively addressed), but the two recognition processes need not be mutually exclusive. PGRP-SA binding to cell wall or PG is boosted by GNBP1 and further enhanced in the presence of PGRP-SD, in the case of specific types of Gram-positive PG (Staphylococci, Streptococci), suggesting that the degree of PG cross-linking may dictate the use of a tripartite complex (SA/SD/GNBP1 in highly linked PG) or the dyad PGRP-SA/GNBP1 in loosely linked PG (Wang et al., 2008). Interestingly, PGRP-SD does not bind PG found in Gram-positive bacteria, so the involvement of this host molecule in sensing these bacteria does not seem to be direct despite the suggestive genetic phenotype (Leone et al., 2008; Wang et al., 2008).

The crystal structure of the N-terminal of GNBP3 gave some insight on how fungal cell wall may be recognized. The binding affinity of this receptor increases with polysaccharide chain length and its binding site is in fact able to discriminate between short- and long-length chains (Mishima et al., 2009). This suggests binding to whole fungal spores.

2.1.4 Ligand activation: The role of Spz

The two strands of sensing microbial molecules from bacteria (PGRP-SA) and fungi (GNBP3) converge to the modular serine protease (ModSP)

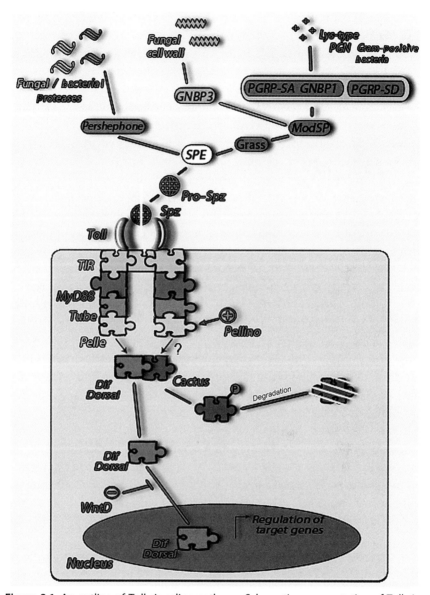

Figure 2.1 An outline of Toll signaling pathway. Schematic representation of Toll signaling in *Drosophila* immunity (see text for details). *Adapted from Kounatidis and Ligoxygakis (2012).* (For color version of this figure, the reader is referred to the online version of this chapter.)

(Buchon et al., 2009), which in turn activates—not directly—the serine protease Grass (Buchon et al., 2009; Kambris et al., 2006). It is an open question whether as a modular protein ModSP may function as a hub for bringing in sensor proteins as well as downstream proteases. The latter, that is, the serine protease immediately downstream of ModSP, is still elusive, but proteases Spirit, Spheroide, and Sphinx1/2 were also identified as necessary for a host responding to both fungi and Gram-positive bacteria (Kambris et al., 2006). Further downstream, the protease Spätzle-processing enzyme (SPE) is the point where pathogen recognition information is integrated also from Psh (triggered by fungal or bacterial proteases). SPE activates the Toll ligand Spz by cleaving its N-terminal prodomain (Jang et al., 2006). Spz binds the receptor through two Spz dimers, each interacting with the N terminus of one Toll molecule, triggering a conformational change in what is now a dimeric Toll receptor, to activate downstream signaling (Gangloff et al., 2008). It is still an open question whether the Spz prodomain is separated from the hydrophobic C-106 domain when cleaved, as has been suggested *in vivo* (Morisato, 2001), or remains attached through disulphide bonds, as seen in biochemical experiments, to be finally displaced when bound to Toll (Weber et al., 2007). Whatever the case, recent data combining computer modeling and experiments in the early embryo indicate an inhibitory role for the prodomain along with the known signaling role of C-106 where the pro-domain may shape the way Spz itself is cleaved and diffused (Haskel-Ittah et al., 2012). Any consequences from this in immunity remain unexplored.

The signaling role of Spz in immunity is an interesting one. Transcriptomic analysis in larvae has shown that following infection, *spz* expression is induced in blood cells significantly more than in fat body, despite the fact that it is expressed in both tissues (Irving et al., 2005). Indeed, tissue-specific knockdown of *spz* in larval blood cells was shown to be important for Toll-dependent AMP induction in fat body (Shia et al., 2009). This indicates that Spz secreted by blood cells induces systemic immunity in the fat body in a manner that parallels mammalian cytokine activity. Spz, however, might not be the only signal, as genetic ablation of blood cells in larvae had a global effect on AMP production in the fat body (Shia et al., 2009). These phenomena were not observed in adults (Charroux & Royet, 2009; Shia et al., 2009).

The discrepancy between the paracrine role of Spz in larvae and the apparent lack of such activity in adults leads to an attractive conceptual framework of how *Drosophila* immunity works during development.

Timeline of development in embryogenesis for two of the major immune responsive tissues (blood cells and fat body) has a lag phase where blood cells appear first (late stage 10), while fat cell precursors appear afterward (late stage 15) (Hoshizaki et al., 1994; Tepass et al., 1994). Interestingly, blood cells rapidly become macrophages (Tepass et al., 1994), whereas fat cell progenitors undergo several rounds of endocrine signaling and maturation leading to mid-larval stage (Hoshizaki et al., 1994). This means that during embryogenesis and early larval development, blood cells are the major immune component. As the fat body matures in the late larva/pupal transition, the major role may be taken over by the mature "remodeled" fat body, and in the emerging adult, it is the latter that carries most functions for the coordination of systemic responses. Nevertheless, recent studies show that blood cells in adults still retain a paracrine role (Clark, Woodcock, Geissmann, Trouillet, & Dionne, 2011).

2.2. Intracellular signaling

Following Spz–Toll interaction, a receptor–adaptor complex that will transmit the signal from the cell surface to the nucleus is formed. This complex comprises the MyD88 protein, which interacts with Toll through their respective Toll/Interleukin-1 receptor domains (Tauszig-Delamasure, Bilak, Capovilla, Hoffmann, & Imler, 2002) at the cell surface (Marek & Kagan, 2012) and connects with Tube (Towb, Sun, & Wasserman, 2009) via death domain contacts that will in turn recruit the kinase Pelle (Towb, Bergmann, & Wasserman, 2001). Recruitment of Tube by MyD88 is a process dependent on the C-terminal phosphoinositide-binding domain found in the latter (Marek & Kagan, 2012). It seems, therefore, that *Drosophila* MyD88 combines the properties of murine MyD88 as well as the phosphoinositide-binding adaptor TIRAP needed for the recruitment of MyD88 to the cell surface (Marek & Kagan, 2012). Tube and Pelle are related to the IRAK family of protein kinases and seem to have been derived from duplication of an ancestral gene (Towb et al., 2009). Thereafter, Tube evolved as an adaptor protein, while Pelle retained the original function. Despite lacking kinase ability, Tube is the *Drosophila* IRAK-4 homologue, while Pelle is the IRAK-1 homologue (Gosu, Basith, Durai, & Choi, 2012; Towb et al., 2009). Pelle will directly or indirectly phosphorylate the IκB homologue Cactus, which is thus targeted for degradation (Belvin, Jin, & Anderson, 1995). Upon Cactus degradation, the NF-κB homologues Dorsal or Dif are free to move to the nucleus and regulate hundreds of target genes (Manfruelli, Reichhart, Steward, Hoffmann, & Lemaitre, 1999; Rutschmann et al., 2000). A positive regulator of the pathway is the RING domain-containing Pellino, acting presumably at the level of Pelle in parallel to mammalian Pellinos that modulate IRAK action through degradation-independent ubiquitination (Haghayeghi, Sarac, Czerniecki, Grosshans, & Schöck, 2010). In contrast, a negative regulator is WntD, which reduces Toll activity by preventing translocation of Dorsal to the nucleus (Gordon, Dionne, Schneider, & Nusse, 2005). In addition, it has recently been shown that endocytosis is paramount for efficient Toll signaling (Huang, Chen, Kunes, Chang, & Maniatis, 2010). A schematic summary of Toll pathway signaling is presented in Figure 2.1.

2.3. Host–pathogen interaction upstream of Toll

One unexplored aspect of the pathogen recognition process is what happens when pathogen surfaces change. This is particularly intriguing for Gram-positive bacteria such as the opportunistic pathogen *Staphylococcus aureus*, which have their cell wall exposed to the environment found within a host. Following an infection, these bacteria need to find ways to evade or reduce recognition by the host in order to survive and potentially proliferate. The cell wall of Gram-positive bacteria consists of an intricate network of PG covalently linked to surface proteins and glycopolymers including Wall Teichoic Acids (WTA), and as we have seen previously, PG is a major target for recognition. It has been proposed that lack of WTA reduces the pathogenicity of *S. aureus*. We asked whether this was due to better recognition of PG. We found that both bacterial recognition and survival of fruit flies infected with bacteria lacking WTA were markedly increased compared to those infected with wild-type *S. aureus* and host survival was now independent of both the Toll pathway and PGRP-SD but dependent on PGRP-SA and GNBP1 (Atilano et al., 2011). This result was quantifiable: a reduction in the amount of WTA resulted in greater binding by host receptors and a higher host survival. Our model is that the presence of WTA limits access to PG and therefore reduces the recognition ability of the host. Bacteria are thus able to increase in numbers and eventually overwhelm the host (Atilano et al., 2011).

We recently found another strategy used by Gram-positive bacteria to avoid recognition in addition to the existence of WTAs, namely, autolysins. The latter are enzymes capable of hydrolyzing PG and have a major role in concealing this inflammatory molecule from *Drosophila* PGRPs. *S. aureus* mutants lacking autolysins are very vulnerable to the fruit fly's immune system in a PGRP-SA-dependent manner (Atilano et al., submitted). This is because autolysins trim the outermost PG fragments and in their absence as PGRP-SA can directly recognize leftover PG extending beyond the external layers of bacterial proteins and polysaccharides. Interestingly, the activity of autolysins is not restricted to the producer cells, but it can also alter the surface of neighboring bacteria, facilitating the survival of the entire population in the infected host (Atilano et al., submitted). Crucially, double *S aureus* mutants lacking both WTAs and autolysin become more sensitive to killing than their single mutant counterparts (Atilano et al., submitted) showing the versatility of strategies that Gram-positive bacteria employ to evade or reduce host recognition.

Finally, gastrointestinal infection of larvae with the fungus *C. albicans* provokes systemic immunity driven by the fat body (Glittenberg, Kounatidis, et al., 2011). This implies a communication between the gut and the fat body through both the host and the pathogen. From the side of the host, Toll signaling as well as Nitric Oxide is required with secreted aspartyl proteases 4 and 6 playing a role from the side of the pathogen, in agreement with mouse models (Glittenberg, Kounatidis, et al., 2011).

3. THE IMMUNE DEFICIENCY PATHWAY

3.1. Recognition and intracellular signaling

It is striking that *Drosophila* does not recognize LPS from Gram-negative bacteria but PG (Kaneko et al., 2004; Leulier et al., 2003). Infection by these bacteria triggers the immune deficiency (IMD) pathway, which shows striking similarities to the ones stimulated by members of the mammalian TNF-receptor super-family. It is assumed that fragments of PG released by these bacteria bind the PGRPs LC or LE at the cell surface or the cytosol, respectively, leading to their multimerization (Kaneko et al., 2006). Whether PGRP-LC/LE can bind whole bacteria is still an open question although it is the tracheal cytotoxin (TCT), a monomeric disaccharide–tetrapeptide fragment of PG, that has been implicated both in cell culture (Kaneko et al., 2004) and *in vivo* (Bosco-Drayon et al., 2012; Neyen, Poidevin, Roussel, & Lemaitre, 2012) as a long range elicitor of immunity. Differential splicing of PGRP-LC domain-encoding exons to a common intracellular domain-encoding exon generates three receptor isoforms, which differ in their PG-binding specificities. Recently, Phi31-mediated recombineering was used to generate fly lines expressing specific isoforms of PGRP-LC, thus assessing the tissue-specific roles of PGRP-LC isoforms and PGRP-LE in the antibacterial response (Neyen et al., 2012). PGRP-LCx had a key role in sensing PG from Gram–negative bacteria or Gram–positive bacilli during systemic infection (Neyen et al., 2012). In addition, PGRP-LCa/x heterodimers had an important contribution to systemic immune response against Gram–negative bacteria through sensing of TCT, whereas PGRP-LCy may have a minor role in antagonizing the immune response (Neyen et al., 2012). Parallel studies have shown that the *Drosophila* gut is regionalized with PGRP-LE being very important in the midgut as a putative intracellular sensor of TCT and PGRP-LC needed upstream where food is broken down mechanically (in the proventriculus; Bosco-Drayon et al., 2012). These results reveal that both PGRP-LC and PGRP-LE contribute to the intestinal immune response, with a predominant role of cytosolic PGRP-LE in the midgut.

Upon recognition, the signal is transduced through a receptor–adaptor complex comprising Imd itself (homologous to the mammalian receptor interacting protein (RIP), minus the kinase domain) (Georgel et al., 2001), dFADD (FAS-associated death domain protein homologue) (Leulier, Vidal, Saigo, Ueda, & Lemaitre, 2002; Naitza et al., 2002), and the caspase-8 homologue Dredd (death-related Ced-3/Nedd2-like protein) (Leulier, Rodriguez, Khush, Abrams, & Lemaitre, 2000). DREDD is K63-linked ubiquitinated by the *Drosophila* Inhibitor of apoptosis-2 (dIAP-2), which acts as an E3-ligase promoting DREDD activation (Meinander et al., 2012). DREDD then cleaves Imd, thus unmasking a domain of interaction with dIAP-2 for further dIAP-2-dependent Ub-conjugation this time on Imd itself (Paquette et al., 2010). Through its RING domain, dIAP-2 ubiquitinates and stabilizes Imd, which then acts like a scaffold for the recruitment of downstream components (Paquette et al., 2010). These components may include the *Drosophila* TGF-beta-activating-kinase 1 (dTAK1), together with adaptor TAK1-associated binding protein 2 (dTAB2) (Kleino et al., 2005; Paquette et al., 2010; Vidal et al., 2001). TAK1/TAB2 play a critical role in activating the *Drosophila* IκB kinase (IKK) complex and also transiently activate the c-Jun-N-terminal kinase (JNK) pathway (Silverman et al., 2003). In this IKK/JNK dichotomy, the IKK complex represents the immune branch of the pathway and phosphorylates the NF-κB transcription factor Relish (Ertürk-Hasdemir et al., 2009). It is probable (but not proved) that Dredd cleaves the inhibitory C-terminal domain of phosphorylated Relish helped by an IKK scaffold (Ertürk-Hasdemir et al., 2009; Stoven et al., 2003), thus liberating the N-terminal portion to move to the nucleus and regulate expression of transcriptional targets such as the AMP gene *diptericin* (*dipt*).

IMD signaling shows an acute phase profile and is rapidly terminated after initial infection (Lemaitre et al., 1997). Negative regulation plays an important role in restricting the response both inside and outside the cell in epithelia and systemic infection. Outside the cell, PGRPs with an amidase activity act to downregulate the pathway following microbial sensing (Paredes, Welchman, Poidevin, & Lemaitre, 2011). Inside the cell, Pirk negatively regulates the receptor PGRP-LC (Aggarwal et al., 2008; Kleino et al., 2008; Lhocine et al., 2008), while deubiquitinases (DUBs) dUSP36 and CYLD inhibit, respectively, Imd itself (Thevenon et al., 2009) and the IKK complex (Tsichritzis et al., 2007). This is presumably achieved by deassembling the transient signaling center generated by the addition of Ub chains, which as in mammals starts with the receptor–adaptor complex (PGRP-LC, Imd, Dredd, dFADD) and continues with the recruitment of TAK1/TAB2 and IKK. Relish also plays a crucial role in limiting the signal through proteosomal degradation of dTAK1 (Park et al., 2004). Nevertheless, it is still unclear how dTAK1 is activated although both Dredd and K63-polyubiquitin chains may be involved (Ertürk-Hasdemir et al., 2009; Meinander et al., 2012). If the latter is true, then an additional DUB to pacify the signal at the level of TAK1/TAB2 may be involved, but this remains to be established. Finally, at the level of AMP induction, the Hox gene *Caudal* is keeping transcription off in the absence of pathogenic infection (Ryu et al., 2008). The above bring forward IMD as an important control point for immune tolerance (Figure 2.2).

3.2. Gut immunity and IMD signaling

IMD has most recently been involved in gut immunity. Anatomically, the *Drosophila* gut can be divided into foregut, midgut, and hindgut. The upper digestive system is used for food uptake and storage, while processing and absorption take place in the mid and posterior regions of the midgut. In this continuous system typical of higher Diptera, some of the meal is completely processed and defecated before some has even entered the digestive section of the midgut. The presence of intestinal stem cells (ISCs) ensures gut homeostasis with the supply of differentiated enterocytes (ECs). A characteristic of ECs is their rapid turnover where apoptotic cells are replaced by the compensatory. Similar to mammals, the Notch, Wingless, platelet-derived growth factor (PDGF), epidermal growth factor (EGF), and insulin receptor pathways have been implicated in the maintenance, proliferation, and/or differentiation of ISCs (see Jiang & Edgar 2011

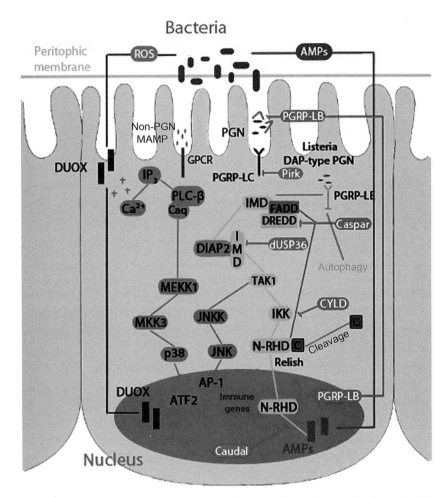

Figure 2.2 An outline of IMD signaling pathway. IMD signaling is involved both in innate immunity and in systemic immunity. Signaling is depicted in the gut to connect IMD with additional epithelial responses such as reactive oxygen species (ROS), the production of which is mediated by the *Drosophila* dual oxidase (dDuox). See text for details. *Adapted from Kounatidis and Ligoxygakis (2012).* (For color version of this figure, the reader is referred to the online version of this chapter.)

for a review). In addition, Hippo signaling is used to restrict stem-cell proliferation in the gut of both *Drosophila* and mammals (Poernbacher, Baumgartner, Marada, Edwards, & Stocker, 2012). Fox and Spradling have settled a controversy in the field by recording the absence of active stem cells but the presence of Wingless-expressing cells within the anterior pylorus, the proliferation of which provides homeostasis following serious damage

(Fox & Spradling, 2009). Gut responses in the gut are highly regionalized as food is mechanically broken in the proventriculus (foregut) and moves posteriorly through the Copper cells region to the posterior midgut. In that sense, there is a shift in PGRP requirement from anterior (side of food entry) to posterior where PGRP-LC is required in the former and PGRP-LE in the latter part of the gut to induce IMD signaling (Bosco-Drayon et al., 2012). PGRP-LE is important for both pathogenic and commensal bacteria. However, in order to maintain immune tolerance, amidase PGRPs have also an important role to play: PGRP-LE activation is inducing amidase PGRPs to hydrolyze PG and remove the stimulus (Bosco-Drayon et al., 2012). Thus, amidase PGRPs dampen any PGRP-LE-mediated response. Removal of one of those amidase PGRPs (PGRP-LB) leads to reduction in life span (Paredes et al., 2011) due to chronic activation of immunity, which can be rescued with concomitant removal of PGRP-LE (Bosco-Drayon et al., 2012).

3.3. Host–pathogen interaction

In addition to recognition avoidance strategies of Gram-positive bacteria, clues on how Gram-negative bacteria may persist in the host have been identified. The *evf* (*Erwinia* virulence factor) gene has been identified as important for persistence in *Erwinia carotovora* infection (Basset, Tzou, Lemaitre, & Boccard, 2003). Transfer of this gene to other Gram-negative bacteria allows them to persist following ingestion indicating that a single gene is able to confer such properties to bacteria (Basset et al., 2003). Consistent with the idea that *evf* somehow perturbs gut physiology, overexpression of *evf* results in increased bacterial accumulation (Basset et al., 2003). In addition, larvae stop feeding and a strong lethality phenotype is observed (Basset et al., 2003). Ingestion by another bacterium, namely, *Pseudomonas entomophila*, is also characterized by cessation of feeding and gut damage (Vodovar et al., 2005). Again, a single gene coding for the metalloproteinase AprA is largely responsible for driving persistence and pathogenesis of this bacterium (Liehl, Blight, Vodovar, Boccard, & Lemaitre, 2006).

4. IMMUNITY AS A SENSOR OF METABOLISM

Given that mounting an immune response is costly in terms of energy, recent studies have examined the possible nexus between immunity and metabolic homeostasis. In this context, the transcription factor FOXO has been found to be central to this connection since in addition to its role as a metabolic regulator, FOXO was recently found to regulate immune homeostasis by transcriptional control of AMPs (Becker et al., 2010).

Upon infection or overexpression, Toll pathway activation was found to suppress insulin signaling by altering the subcellular localization of FOXO, which then accumulates in the nucleus blocking insulin and growth (DiAngelo, Bland, Bambina, Cherry, & Birnbaum, 2009). It has been found that *Mycobacterium* infection leads also to increase in FOXO activity, which results in infection-induced wasting (Dionne, Pham, Shirasu-Hiza, & Schneider, 2006). Finally, a separate challenge, namely, localized DNA damage, activates the immune response, leading in turn to blockage of transcription of *Drosophila* insulin-like peptides in the median neurosecretory cells in the brain (Karpac, Younger, & Jasper, 2011). This in turn results in increased survival and underlines the importance of the interaction between immunity and growth homeostasis. It also brings forward the fact that genetic screens using survival following infection as an assay could still uncover very important aspects of interorgan regulation in host defense.

5. CONCLUDING REMARKS—FUTURE DIRECTIONS

Studies of *Drosophila* immunity have put innate immunity under an evolutionary light, and the model has been used not only as a roadmap in mammals but also as a blueprint for what could happen in other dipteran insect species of medical importance (Waterhouse et al., 2007). In this context, studies of *Drosophila* with natural gut-dwelling trypanosomid parasites are of great importance especially for those vectors of neglected tropical disease difficult to culture and study in the lab (tsetse flies, sand flies, etc.). At the same time, the host–pathogen interaction aspect at the molecular level is the one that has not been systematically explored. We learned a lot about the host reaction but do not know enough about how this reaction is altered when the pathogen changes. In addition, we do not have any information of how recognition is achieved *in vivo* and in real time especially for Gram-positive bacteria where secreted components are involved in pathogen sensing. Is there a complex formed on bacterial surfaces? Does the modular ModSP protease function as a platform for complex formation? We have seen that there is a PGRP-SA-dependent but Toll-independent aspect of countering Gram-positive bacteria (Atilano et al., 2011). It is an intriguing question what distinguishes the need or not for Toll and the nature of the Toll-independent killing mechanism.

Finally, unlike another invertebrate model (the worm *C. elegans*), *Drosophila* has not been repeatedly used as a high-throughput system to study virulence of human pathogens or antimicrobial agents (Kirienko et al., 2013). Nevertheless, feasibility studies have been performed (Chamilos et al., 2008, 2006; Glittenberg, Kounatidis, et al., 2011b; Glittenberg, Silas, MacCallum, Gow, & Ligoxygakis, 2011a; Lamaris et al., 2009), and as pathogen deletion mutant libraries become available, *Drosophila* will certainly be the model of choice for such screens. The extensive conservation between *Drosophila* host defense and mammalian innate immunity will be able to largely reduce the use of animals and focus studies on interesting candidates since there is parallel between virulence in flies and mice (Glittenberg, Silas, et al., 2011). In the future, studies into these issues and beyond will generate exciting biology and reveal new and surprising aspects in the evolution and regulation of immunity. Already, one such underappreciated aspect is behavior (Kacsoh, Lynch, Mortimer, & Schlenke, 2013). When flies see parasitic wasps, they switch to laying eggs in alcohol-laden food sources that protect hatched larvae from infection. This change in oviposition behavior, mediated by neuropeptide F, is retained long after wasps are removed (Kacsoh et al., 2013).

ACKNOWLEDGMENTS

The author would like to thank Dr. Ilias Kounatidis for drawing the figures. Our work is supported by the European Research Council (Starter Grant 2012 "Droso-Parasite"), the Biological and Biotechnological Science Research Council (project Grant BB/K003569/1), and the National Committee for the Reduction, Refinement and Replacement of the Use of Animals, UK (NC3Rs).

REFERENCES

Aggarwal, K., Rus, F., Vriesema-Magnuson, C., Ertürk-Hasdemir, D., Paquette, N., & Silverman, N. (2008). Rudra interrupts receptor signaling complexes to negatively regulate the IMD pathway. *PLoS Pathogens, 4*(8), e1000120. http://dx.doi.org/10.1371/journal.ppat.1000120.

Atilano, M. L., Yates, J., Glittenberg, M., Filipe, S. R., & Ligoxygakis, P. (2011). Wall teichoic acids of Staphylococcus aureus limit recognition by the drosophila peptidoglycan recognition protein-SA to promote pathogenicity. *PLoS Pathogens, 7*(12), e1002421. http://dx.doi.org/10.1371/journal.ppat.1002421.

Audry, M., Ciancanelli, M., Yang, K., Cobat, A., Chang, H.-H., Sancho-Shimizu, V., et al. (2011). NEMO is a key component of NF-κB- and IRF-3-dependent TLR3-mediated immunity to herpes simplex virus. *The Journal of Allergy and Clinical Immunology, 128*(3), 610–617. http://dx.doi.org/10.1016/j.jaci.2011.04.059, e1–4.

Avet-Rochex, A., Boyer, K., Polesello, C., Gobert, V., Osman, D., Roch, F., et al. (2010). An in vivo RNA interference screen identifies gene networks controlling Drosophila melanogaster blood cell homeostasis. *BMC Developmental Biology*, *10*(1), 65. http://dx. doi.org/10.1186/1471-213X-10-65.

Basset, A., Tzou, P., Lemaitre, B., & Boccard, F. (2003). A single gene that promotes inter-action of a phytopathogenic bacterium with its insect vector, Drosophila melanogaster. *EMBO Reports*, *4*(2), 205–209. http://dx.doi.org/10.1038/sj.embor.embor730.

Becker, T., Loch, G., Beyer, M., Zinke, I., Aschenbrenner, A. C., Carrera, P., et al. (2010). FOXO-dependent regulation of innate immune homeostasis. *Nature*, *463*(7279), 369–373. http://dx.doi.org/10.1038/nature08698.

Belvin, M. P., Jin, Y., & Anderson, K. V. (1995). *Cactus protein degradation mediates Drosophila dorsal-ventral signaling. Genes & Development*, *9*(7), 783–793. Retrieved from, http://www.ncbi.nlm.nih.gov/pubmed/7705656.

Benmimoun, B., Polesello, C., Waltzer, L., & Haenlin, M. (2012). Dual role for Insulin/TOR signaling in the control of hematopoietic progenitor maintenance in Drosophila. *Development (Cambridge, England)*, *139*(10), 1713–1717. http://dx.doi.org/10.1242/dev.080259.

Bischoff, V., Vignal, C., Boneca, I. G., Michel, T., Hoffmann, J. A., & Royet, J. (2004). Function of the drosophila pattern-recognition receptor PGRP-SD in the detection of Gram-positive bacteria. *Nature Immunology*, *5*(11), 1175 1180. http://dx.doi.org/10.1038/ni1123.

Bosco-Drayon, V., Poidevin, M., Boneca, I. G., Narbonne-Reveau, K., Royet, J., & Charroux, B. (2012). Peptidoglycan sensing by the receptor PGRP-LE in the Drosophila gut induces immune responses to infectious bacteria and tolerance to microbiota. *Cell Host & Microbe*, *12*(2), 153–165. http://dx.doi.org/10.1016/j.chom.2012.06.002.

Bousfiha, A., Picard, C., Boisson-Dupuis, S., Zhang, S.-Y., Bustamante, J., Puel, A., et al. (2010). Primary immunodeficiencies of protective immunity to primary infections. *Clinical Immunology (Orlando, Fla.)*, *135*(2), 204–209. http://dx.doi.org/10.1016/j.clim.2010.02.001.

Buchon, N., Poidevin, M., Kwon, H.-M., Guillou, A., Sottas, V., Lee, B.-L., et al. (2009). A single modular serine protease integrates signals from pattern-recognition receptors upstream of the Drosophila Toll pathway. *Proceedings of the National Academy of Sciences of the United States of America*, *106*(30), 12442–12447. http://dx.doi.org/10.1073/pnas.0901924106.

Casanova, J.-L., Abel, L., & Quintana-Murci, L. (2011). Human TLRs and IL-1Rs in host defense: Natural insights from evolutionary, epidemiological, and clinical genetics. *Annual Review of Immunology*, *29*(2010), 447–491. http://dx.doi.org/10.1146/annurev-immunol-030409-101335.

Chamilos, G., Lewis, R. E., Hu, J., Xiao, L., Zal, T., Gilliet, M., et al. (2008). Drosophila melanogaster as a model host to dissect the immunopathogenesis of zygomycosis. *Proceedings of the National Academy of Sciences of the United States of America*, *105*(27), 9367–9372. http://dx.doi.org/10.1073/pnas.0709578105.

Chamilos, G., Lionakis, M. S., Lewis, R. E., Lopez-Ribot, J. L., Saville, S. P., Albert, N. D., et al. (2006). Drosophila melanogaster as a facile model for large-scale studies of virulence mechanisms and antifungal drug efficacy in Candida species. *The Journal of Infectious Diseases*, *193*(7), 1014–1022. http://dx.doi.org/10.1086/500950.

Charroux, B., & Royet, J. (2009). Elimination of plasmatocytes by targeted apoptosis reveals their role in multiple aspects of the Drosophila immune response. *Proceedings of the National Academy of Sciences of the United States of America*, *106*(24), 9797–9802. http://dx.doi.org/10.1073/pnas.0903971106.

Clark, R. I., Woodcock, K. J., Geissmann, F., Trouillet, C., & Dionne, M. S. (2011). Multiple TGF-β superfamily signals modulate the adult Drosophila immune response. *Current Biology: CB*, *21*(19), 1672–1677. http://dx.doi.org/10.1016/j.cub.2011.08.048.

Crozatier, M., & Vincent, A. (2011). Drosophila: A model for studying genetic and molecular aspects of haematopoiesis and associated leukaemias. *Disease Models & Mechanisms*, *4*(4), 439–445. http://dx.doi.org/10.1242/dmm.007351.

DeLotto, Y., & DeLotto, R. (1998). *Proteolytic processing of the Drosophila Spätzle protein by easter generates a dimeric NGF-like molecule with ventralising activity. Mechanisms of Development,* 72(1–2), 141–148. Retrieved from, http://www.ncbi.nlm.nih.gov/pubmed/9533958.

DiAngelo, J. R., Bland, M. L., Bambina, S., Cherry, S., & Birnbaum, M. J. (2009). The immune response attenuates growth and nutrient storage in Drosophila by reducing insulin signaling. *Proceedings of the National Academy of Sciences of the United States of America,* 106(49), 20853–20858. http://dx.doi.org/10.1073/pnas.0906749106.

Dionne, M. S., Pham, L. N., Shirasu-Hiza, M., & Schneider, D. S. (2006). Akt and FOXO dysregulation contribute to infection-induced wasting in Drosophila. *Current Biology: CB,* 16(20), 1977–1985. http://dx.doi.org/10.1016/j.cub.2006.08.052.

Dong, Y., Cirimotich, C. M., Pike, A., Chandra, R., & Dimopoulos, G. (2012). Anopheles NF-κB-regulated splicing factors direct pathogen-specific repertoires of the hypervariable pattern recognition receptor AgDscam. *Cell Host & Microbe,* 12(4), 521–530. http://dx.doi.org/10.1016/j.chom.2012.09.004.

El Chamy, L., Leclerc, V., Caldelari, I., & Reichhart, J.-M. (2008). Sensing of "danger signals" and pathogen-associated molecular patterns defines binary signaling pathways "upstream" of Toll. *Nature Immunology,* 9(10), 1165–1170. http://dx.doi.org/10.1038/ni.1643.

Ertürk-Hasdemir, D., Broemer, M., Leulier, F., Lane, W. S., Paquette, N., Hwang, D., et al. (2009). Two roles for the Drosophila IKK complex in the activation of Relish and the induction of antimicrobial peptide genes. *Proceedings of the National Academy of Sciences of the United States of America,* 106(24), 9779–9784. http://dx.doi.org/10.1073/pnas.0812022106.

Fox, D. T., & Spradling, A. C. (2009). The Drosophila hindgut lacks constitutively active adult stem cells but proliferates in response to tissue damage. *Cell Stem Cell,* 5(3), 290–297. http://dx.doi.org/10.1016/j.stem.2009.06.003.

Gangloff, M., Murali, A., Xiong, J., Arnot, C. J., Weber, A. N., Sandercock, A. M., et al. (2008). Structural insight into the mechanism of activation of the Toll receptor by the dimeric ligand Spätzle. *The Journal of Biological Chemistry,* 283(21), 14629–14635. http://dx.doi.org/10.1074/jbc.M800112200.

Georgel, P., Naitza, S., Kappler, C., Ferrandon, D., Zachary, D., Swimmer, C., et al. (2001). *Drosophila immune deficiency (IMD) is a death domain protein that activates antibacterial defense and can promote apoptosis. Developmental Cell,* 1(4), 503–514. Retrieved from, http://www.ncbi.nlm.nih.gov/pubmed/11703941.

Glittenberg, M. T., Kounatidis, I., Christensen, D., Kostov, M., Kimber, S., Roberts, I., et al. (2011b). Pathogen and host factors are needed to provoke a systemic host response to gastrointestinal infection of Drosophila larvae by Candida albicans. *Disease Models & Mechanisms,* 4(4), 515–525. http://dx.doi.org/10.1242/dmm.006627.

Glittenberg, M. T., Silas, S., MacCallum, D. M., Gow, N. A. R., & Ligoxygakis, P. (2011a). Wild-type Drosophila melanogaster as an alternative model system for investigating the pathogenicity of Candida albicans. *Disease Models & Mechanisms,* 4(4), 504–514. http://dx.doi.org/10.1242/dmm.006619.

Gobert, V., Gottar, M., Matskevich, A. A., Rutschmann, S., Royet, J., Belvin, M., et al. (2003). Dual activation of the Drosophila toll pathway by two pattern recognition receptors. *Science (New York, N.Y.),* 302(5653), 2126–2130. http://dx.doi.org/10.1126/science.1085432.

Gordon, M. D., Dionne, M. S., Schneider, D. S., & Nusse, R. (2005). WntD is a feedback inhibitor of Dorsal/NF-kappaB in Drosophila development and immunity. *Nature,* 437(7059), 746–749. http://dx.doi.org/10.1038/nature04073.

Gosu, V., Basith, S., Durai, P., & Choi, S. (2012). Molecular evolution and structural features of IRAK family members. *PLoS One,* 7(11), e49771. http://dx.doi.org/10.1371/journal.pone.0049771.

Gottar, M., Gobert, V., Matskevich, A. A., Reichhart, J.-M., Wang, C., Butt, T. M., et al. (2006). Dual detection of fungal infections in Drosophila via recognition of glucans and sensing of virulence factors. *Cell*, *127*(7), 1425–1437. http://dx.doi.org/10.1016/j.cell.2006.10.046.

Haghayeghi, A., Sarac, A., Czerniecki, S., Grosshans, J., & Schöck, F. (2010). Pellino enhances innate immunity in Drosophila. *Mechanisms of Development*, *127*(5–6), 301–307. http://dx.doi.org/10.1016/j.mod.2010.01.004.

Haine, E. R., Moret, Y., Siva-Jothy, M. T., & Rolff, J. (2008). Antimicrobial defense and persistent infection in insects. *Science (New York, N.Y.)*, *322*(5905), 1257–1259. http://dx.doi.org/10.1126/science.1165265.

Haskel-Ittah, M., Ben-Zvi, D., Branski-Arieli, M., Schejter, E. D., Shilo, B.-Z., & Barkai, N. (2012). Self-organized shuttling: Generating sharp dorsoventral polarity in the early Drosophila embryo. *Cell*, *150*(5), 1016–1028. http://dx.doi.org/10.1016/j.cell.2012.06.044.

Hetru, C., Troxler, L., & Hoffmann, J. A. (2003). Drosophila melanogaster antimicrobial defense. *The Journal of Infectious Diseases*, *187*(Suppl. 2), S327–S334. http://dx.doi.org/10.1086/374758.

Hoshino, K., Takeuchi, O., Kawai, T., Sanjo, H., Ogawa, T., Takeda, Y., et al. (1999). *Cutting edge: Toll-like receptor 4 (TLR4)-deficient mice are hyporesponsive to lipopolysaccharide: Evidence for TLR4 as the Lps gene product. Journal of Immunology (Baltimore, Md.: 1950)*, *162*(7), 3749–3752. Retrieved from, http://www.ncbi.nlm.nih.gov/pubmed/10201887.

Hoshizaki, D. K., Blackburn, T., Price, C., Ghosh, M., Miles, K., Ragucci, M., et al. (1994). *Embryonic fat-cell lineage in Drosophila melanogaster. Development (Cambridge, England)*, *120*(9), 2489–2499. Retrieved from, http://www.ncbi.nlm.nih.gov/pubmed/7956826.

Huang, H.-R., Chen, Z. J., Kunes, S., Chang, G.-D., & Maniatis, T. (2010). Endocytic pathway is required for Drosophila Toll innate immune signaling. *Proceedings of the National Academy of Sciences of the United States of America*, *107*(18), 8322–8327. http://dx.doi.org/10.1073/pnas.1004031107.

Irving, P., Troxler, L., Heuer, T. S., Belvin, M., Kopczynski, C., Reichhart, J. M., et al. (2001). A genome-wide analysis of immune responses in Drosophila. *Proceedings of the National Academy of Sciences of the United States of America*, *98*(26), 15119–15124. http://dx.doi.org/10.1073/pnas.261573998.

Irving, P., Ubeda, J.-M., Doucet, D., Troxler, L., Lagueux, M., Zachary, D., et al. (2005). New insights into Drosophila larval haemocyte functions through genome-wide analysis. *Cellular Microbiology*, *7*(3), 335–350. http://dx.doi.org/10.1111/j.1462-5822.2004.00462.x.

Janeway, C. A. (1989). *Approaching the asymptote? Evolution and revolution in immunology. Cold Spring Harbor Symposia on Quantitative Biology*, *54*(Pt 1), 1–13. Retrieved from, http://www.ncbi.nlm.nih.gov/pubmed/2700931.

Janeway, C. A., & Medzhitov, R. (2002). Innate immune recognition. *Annual Review of Immunology*, *20*(2), 197–216. http://dx.doi.org/10.1146/annurev.immunol.20.083001.084359.

Jang, I.-H., Chosa, N., Kim, S.-H., Nam, H.-J., Lemaitre, B., Ochiai, M., et al. (2006). A Spätzle-processing enzyme required for toll signaling activation in Drosophila innate immunity. *Developmental Cell*, *10*(1), 45–55. http://dx.doi.org/10.1016/j.devcel.2005.11.013.

Jiang, H., & Edgar, B. A. (2011). Intestinal stem cells in the adult Drosophila midgut. *Experimental Cell Research*, *317*(19), 2780–2788. http://dx.doi.org/10.1016/j.yexcr.2011.07.020.

Jiang, H., Vilcinskas, A., & Kanost, M. R. (2010). *Immunity in lepidopteran insects. Advances in Experimental Medicine and Biology*, *708*, 181–204. Retrieved from, http://www.ncbi.nlm.nih.gov/pubmed/21528699.

Kacsoh, B. Z., Lynch, Z. R., Mortimer, N. T., & Schlenke, T. A. (2013). Fruit flies medicate offspring after seeing parasites. *Science (New York, N.Y.)*, *339*(6122), 947–950. http://dx. doi.org/10.1126/science.1229625.

Kambris, Z., Brun, S., Jang, I.-H., Nam, H.-J., Romeo, Y., Takahashi, K., et al. (2006). Drosophila immunity: A large-scale in vivo RNAi screen identifies five serine proteases required for Toll activation. *Current Biology: CB*, *16*(8), 808–813. http://dx.doi.org/ 10.1016/j.cub.2006.03.020.

Kaneko, T., Goldman, W. E., Mellroth, P., Steiner, H., Fukase, K., Kusumoto, S., et al. (2004). *Monomeric and polymeric gram-negative peptidoglycan but not purified LPS stimulate the Drosophila IMD pathway. Immunity*, *20*(5), 637–649. Retrieved from, http://www. ncbi.nlm.nih.gov/pubmed/15142531.

Kaneko, T., Yano, T., Aggarwal, K., Lim, J.-H., Ueda, K., Oshima, Y., et al. (2006). PGRP-LC and PGRP-LE have essential yet distinct functions in the drosophila immune response to monomeric DAP-type peptidoglycan. *Nature Immunology*, *7*(7), 715–723. http://dx.doi.org/10.1038/ni1356.

Karpac, J., Younger, A., & Jasper, H. (2011). Dynamic coordination of innate immune signaling and insulin signaling regulates systemic responses to localized DNA damage. *Developmental Cell*, *20*(6), 841–854. http://dx.doi.org/10.1016/j.devcel.2011.05.011.

Kirienko, N. V., Kirienko, D. R., Larkins-Ford, J., Wählby, C., Ruvkun, G., & Ausubel, F. M. (2013 Apr 17). Pseudomonas aeruginosa disrupts Caenorhabditis elegans iron homeostasis, causing a hypoxic response and death. *Cell Host Microbe*, *13*(4), 406–416. http://dx.doi.org/10.1016/j.chom.2013.03.003.

Kleino, A., Myllymäki, H., Kallio, J., Vanha-aho, L.-M., Oksanen, K., Ulvila, J., et al. (2008). *Pirk is a negative regulator of the Drosophila Imd pathway. Journal of Immunology (Baltimore, Md.: 1950)*, *180*(8), 5413–5422. Retrieved from, http://www.ncbi.nlm. nih.gov/pubmed/18390723.

Kleino, A., Valanne, S., Ulvila, J., Kallio, J., Myllymäki, H., Enwald, H., et al. (2005). Inhibitor of apoptosis 2 and TAK1-binding protein are components of the Drosophila Imd pathway. *The EMBO Journal*, *24*(19), 3423–3434. http://dx.doi.org/10.1038/sj.emboj.7600807.

Kounatidis, I., & Ligoxygakis, P. (2012). Drosophila as a model system to unravel the layers of innate immunity to infection. *Open Biology*, *2*(5), 120075. http://dx.doi.org/10.1098/ rsob.120075.

Ku, C.-L., Von Bernuth, H., Picard, C., Zhang, S.-Y., Chang, H.-H., Yang, K., et al. (2007). Selective predisposition to bacterial infections in IRAK-4-deficient children: IRAK-4-dependent TLRs are otherwise redundant in protective immunity. *The Journal of Experimental Medicine*, *204*(10), 2407–2422. http://dx.doi.org/10.1084/jem.20070628.

Lamaris, G. A., Ben-Ami, R., Lewis, R. E., Chamilos, G., Samonis, G., & Kontoyiannis, D. P. (2009). Increased virulence of Zygomycetes organisms following exposure to voriconazole: A study involving fly and murine models of zygomycosis. *The Journal of Infectious Diseases*, *199*(9), 1399–1406. http://dx.doi.org/10.1086/597615.

Lemaitre, B., Nicolas, E., Michaut, L., Reichhart, J. M., & Hoffmann, J. A. (1996). *The dorsoventral regulatory gene cassette spätzle/Toll/cactus controls the potent antifungal response in Drosophila adults. Cell*, *86*(6), 973–983. Retrieved from, http://www.ncbi.nlm.nih.gov/ pubmed/22611248.

Lemaitre, B., Reichhart, J. M., & Hoffmann, J. A. (1997). Drosophila host defense: *Differential induction of antimicrobial peptide genes after infection by various classes of microorganisms. Proceedings of the National Academy of Sciences of the United States of America*, *94*(26), 14614–14619. Retrieved from, http://www.pubmedcentral.nih.gov/articlerender. fcgi?artid=25070&tool=pmcentrez&rendertype=abstract.

LeMosy, E. K., Hong, C. C., & Hashimoto, C. (1999). *Signal transduction by a protease cascade. Trends in Cell Biology*, *9*(3), 102–107. Retrieved from, http://www.ncbi.nlm.nih.gov/ pubmed/10201075.

Leone, P., Bischoff, V., Kellenberger, C., Hetru, C., Royet, J., & Roussel, A. (2008). Crystal structure of Drosophila PGRP-SD suggests binding to DAP-type but not lysine-type peptidoglycan. *Molecular Immunology*, *45*(9), 2521–2530. http://dx.doi.org/10.1016/j.molimm.2008.01.015.

Leulier, F., Parquet, C., Pili-Floury, S., Ryu, J.-H., Caroff, M., Lee, W.-J., et al. (2003). The Drosophila immune system detects bacteria through specific peptidoglycan recognition. *Nature Immunology*, *4*(5), 478–484. http://dx.doi.org/10.1038/ni922.

Leulier, F., Rodriguez, A., Khush, R. S., Abrams, J. M., & Lemaitre, B. (2000). The Drosophila caspase Dredd is required to resist gram-negative bacterial infection. *EMBO Reports*, *1*(4), 353–358. http://dx.doi.org/10.1093/embo-reports/kvd073.

Leulier, F., Vidal, S., Saigo, K., Ueda, R., & Lemaitre, B. (2002). *Inducible expression of double-stranded RNA reveals a role for dFADD in the regulation of the antibacterial response in Drosophila adults. Current Biology: CB*, *12*(12), 996–1000. Retrieved from, http://www.ncbi.nlm.nih.gov/pubmed/12123572.

Levashina, E. A., Langley, E., Green, C., Gubb, D., Ashburner, M., Hoffmann, J. A., et al. (1999). *Constitutive activation of toll-mediated antifungal defense in serpin-deficient Drosophila. Science (New York, N.Y.)*, *285*(5435), 1917–1919. Retrieved from, http://www.ncbi.nlm.nih.gov/pubmed/104893720.

Lhocine, N., Ribeiro, P. S., Buchon, N., Wepf, A., Wilson, R., Tenev, T., et al. (2008). PIMS modulates immune tolerance by negatively regulating Drosophila innate immune signaling. *Cell Host & Microbe*, *4*(2), 147–158. http://dx.doi.org/10.1016/j.chom.2008.07.004.

Liehl, P., Blight, M., Vodovar, N., Boccard, F., & Lemaitre, B. (2006). Prevalence of local immune response against oral infection in a Drosophila/Pseudomonas infection model. *PLoS Pathogens*, *2*(6), e56. http://dx.doi.org/10.1371/journal.ppat.0020056.

Ligoxygakis, P., Pelte, N., Hoffmann, J. A., & Reichhart, J.-M. (2002a). Activation of Drosophila Toll during fungal infection by a blood serine protease. *Science (New York, N.Y.)*, *297*(5578), 114–116. http://dx.doi.org/10.1126/science.1072391.

Ligoxygakis, P., Pelte, N., Ji, C., Leclerc, V., Duvic, B., Belvin, M., et al. (2002b). *A serpin mutant links Toll activation to melanization in the host defence of Drosophila. The EMBO Journal*, *21*(23), 6330–6337. Retrieved from, http://www.pubmedcentral.nih.gov/articlerender.fcgi?artid=136964&tool=pmcentrez&rendertype=abstract.

Ma, C. (2000). A beta 1,3-glucan recognition protein from an insect, Manduca sexta, agglutinates microorganisms and activates the phenoloxidase cascade. *Journal of Biological Chemistry*, *275*(11), 7505–7514. http://dx.doi.org/10.1074/jbc.275.11.7505.

Manfruelli, P., Reichhart, J. M., Steward, R., Hoffmann, J. A., & Lemaitre, B. (1999). A mosaic analysis in Drosophila fat body cells of the control of antimicrobial peptide genes by the Rel proteins Dorsal and DIF. *The EMBO Journal*, *18*(12), 3380–3391. http://dx.doi.org/10.1093/emboj/18.12.3380.

Marek, L. R., & Kagan, J. C. (2012). Phosphoinositide binding by the Toll adaptor dMyD88 controls antibacterial responses in Drosophila. *Immunity*, *36*(4), 612–622. http://dx.doi.org/10.1016/j.immuni.2012.01.019.

Medzhitov, R. (2009). Approaching the asymptote: 20 years later. *Immunity*, *30*(6), 766–775. http://dx.doi.org/10.1016/j.immuni.2009.06.004.

Medzhitov, R., Preston-Hurlburt, P., & Janeway, C. A. (1997). A human homologue of the Drosophila Toll protein signals activation of adaptive immunity. *Nature*, *388*(6640), 394–397. http://dx.doi.org/10.1038/41131.

Meinander, A., Runchel, C., Tenev, T., Chen, L., Kim, C.-H., Ribeiro, P. S., et al. (2012). Ubiquitylation of the initiator caspase DREDD is required for innate immune signalling. *The EMBO Journal*, *31*(12), 2770–2783. http://dx.doi.org/10.1038/emboj.2012.121.

Michel, T., Reichhart, J. M., Hoffmann, J. A., & Royet, J. (2001). Drosophila Toll is activated by Gram-positive bacteria through a circulating peptidoglycan recognition protein. *Nature, 414*(6865), 756–759. http://dx.doi.org/10.1038/414756a.

Mishima, Y., Quintin, J., Aimanianda, V., Kellenberger, C., Coste, F., Clavaud, C., et al. (2009). The N-terminal domain of Drosophila Gram-negative binding protein 3 (GNBP3) defines a novel family of fungal pattern recognition receptors. *The Journal of Biological Chemistry, 284*(42), 28687–28697. http://dx.doi.org/10.1074/jbc.M109. 034587.

Morisato, D. (2001). *Spätzle regulates the shape of the Dorsal gradient in the Drosophila embryo.* *Development (Cambridge, England), 128*(12), 2309–2319. Retrieved from, http://www. ncbi.nlm.nih.gov/pubmed/11493550.

Naitza, S., Rossé, C., Kappler, C., Georgel, P., Belvin, M., Gubb, D., et al. (2002). The *Drosophila immune defense against gram-negative infection requires the death protein dFADD.* *Immunity, 17*(5), 575–581. Retrieved from, http://www.ncbi.nlm.nih.gov/pubmed/ 12433364.

Nappi, A., Poirié, M., & Carton, Y. (2009). The role of melanization and cytotoxic by-products in the cellular immune responses of Drosophila against parasitic wasps. *Advances in Parasitology, 70,* 99–121. http://dx.doi.org/10.1016/S0065-308X(09) 70004-1.

Neyen, C., Poidevin, M., Roussel, A., & Lemaitre, B. (2012). Tissue- and ligand-specific sensing of gram-negative infection in Drosophila by PGRP-LC isoforms and PGRP-LE. *Journal of Immunology (Baltimore, Md.: 1950), 189*(4), 1886–1897. http://dx.doi. org/10.4049/jimmunol.1201022.

Ochiai, M. (2000). A pattern-recognition protein for beta -1,3-glucan. The binding domain and the cDNA cloning of beta -1,3-glucan recognition protein from the silkworm, Bombyx mori. *Journal of Biological Chemistry, 275*(7), 4995–5002. http://dx.doi.org/ 10.1074/jbc.275.7.4995.

Paquette, N., Broemer, M., Aggarwal, K., Chen, L., Husson, M., Ertürk-Hasdemir, D., et al. (2010). Caspase-mediated cleavage, IAP binding, and ubiquitination: Linking three mechanisms crucial for Drosophila NF-kappaB signaling. *Molecular Cell, 37*(2), 172–182. http://dx.doi.org/10.1016/j.molcel.2009.12.036.

Paredes, J. C., Welchman, D. P., Poidevin, M., & Lemaitre, B. (2011). Negative regulation by amidase PGRPs shapes the Drosophila antibacterial response and protects the fly from innocuous infection. *Immunity, 35*(5), 770–779. http://dx.doi.org/10.1016/j.immuni. 2011.09.018.

Park, J. M., Brady, H., Ruocco, M. G., Sun, H., Williams, D., Lee, S. J., et al. (2004). Targeting of TAK1 by the NF-kappa B protein Relish regulates the JNK-mediated immune response in Drosophila. *Genes & Development, 18*(5), 584–594. http://dx.doi. org/10.1101/gad.1168104.

Pennetier, D., Oyallon, J., Morin-Poulard, I., Dejean, S., Vincent, A., & Crozatier, M. (2012). Size control of the Drosophila hematopoietic niche by bone morphogenetic protein signaling reveals parallels with mammals. *Proceedings of the National Academy of Sciences of the United States of America, 109*(9), 3389–3394. http://dx.doi.org/10.1073/ pnas.1109407109.

Poernbacher, I., Baumgartner, R., Marada, S. K., Edwards, K., & Stocker, H. (2012). Drosophila Pez acts in Hippo signaling to restrict intestinal stem cell proliferation. *Current Biology: CB, 22*(5), 389–396. http://dx.doi.org/10.1016/j.cub.2012.01.019.

Poltorak, A., He, X., Smirnova, I., Liu, M. Y., Van Huffel, C., Du, X., et al. (1998). *Defective LPS signaling in C3H/HeJ and C57BL/10ScCr mice: Mutations in Tlr4 gene. Science (New York, N.Y.), 282*(5396), 2085–2088. Retrieved from, http://www.ncbi.nlm.nih.gov/ pubmed/9851930.

Rizki, T. (1978). Fat body. In M. Ashburner & T. Wright (Eds.), *The genetics and biology of Drosophila* (pp. 561–599). New York: Academic.

Rutschmann, S., Jung, A. C., Hetru, C., Reichhart, J. M., Hoffmann, J. A., & Ferrandon, D. (2000). *The Rel protein DIF mediates the antifungal but not the antibacterial host defense in Drosophila*. Immunity, *12*(5), 569–580. Retrieved from, http://www.ncbi.nlm.nih.gov/pubmed/10843389.

Ryu, J.-H., Kim, S.-H., Lee, H.-Y., Bai, J. Y., Nam, Y.-D., Bae, J.-W., et al. (2008). Innate immune homeostasis by the homeobox gene caudal and commensal-gut mutualism in Drosophila. *Science (New York, N.Y.)*, *319*(5864), 777–782. http://dx.doi.org/10.1126/science.1149357.

Shia, A. K. H., Glittenberg, M., Thompson, G., Weber, A. N., Reichhart, J.-M., & Ligoxygakis, P. (2009). Toll-dependent antimicrobial responses in Drosophila larval fat body require Spätzle secreted by haemocytes. *Journal of Cell Science*, *122*(Pt 24), 4505–4515. http://dx.doi.org/10.1242/jcs.049155.

Silverman, N., Zhou, R., Erlich, R. L., Hunter, M., Bernstein, E., Schneider, D., et al. (2003). Immune activation of NF-kappaB and JNK requires Drosophila TAK1. *The Journal of Biological Chemistry*, *278*(49), 48928–48934. http://dx.doi.org/10.1074/jbc.M304802200.

Steward, R. (1987). *Dorsal, an embryonic polarity gene in Drosophila, is homologous to the vertebrate proto-oncogene, c-rel*. Science (New York, N.Y.), *238*(4827), 692–694. Retrieved from, http://www.ncbi.nlm.nih.gov/pubmed/3118464.

Stoven, S., Silverman, N., Junell, A., Hedengren-Olcott, M., Erturk, D., Engstrom, Y., et al. (2003). Caspase-mediated processing of the Drosophila NF-kappaB factor Relish. *Proceedings of the National Academy of Sciences of the United States of America*, *100*(10), 5991–5996. http://dx.doi.org/10.1073/pnas.1035902100.

Tauszig-Delamasure, S., Bilak, H., Capovilla, M., Hoffmann, J. A., & Imler, J.-L. (2002). Drosophila MyD88 is required for the response to fungal and Gram-positive bacterial infections. *Nature Immunology*, *3*(1), 91–97. http://dx.doi.org/10.1038/ni747.

Tepass, U., Fessler, L. I., Aziz, A., & Hartenstein, V. (1994). *Embryonic origin of hemocytes and their relationship to cell death in Drosophila*. Development (Cambridge, England), *120*(7), 1829–1837. Retrieved from, http://www.ncbi.nlm.nih.gov/pubmed/7924990.

Thevenon, D., Engel, E., Avet-Rochex, A., Gottar, M., Bergeret, E., Tricoire, H., et al. (2009). The Drosophila ubiquitin-specific protease dUSP36/Scny targets IMD to prevent constitutive immune signaling. *Cell Host & Microbe*, *6*(4), 309–320. http://dx.doi.org/10.1016/j.chom.2009.09.007.

Tokusumi, Y., Tokusumi, T., Shoue, D. A., & Schulz, R. A. (2012). Gene regulatory networks controlling hematopoietic progenitor niche cell production and differentiation in the Drosophila lymph gland. *PLoS One*, *7*(7), e41604. http://dx.doi.org/10.1371/journal.pone.0041604.

Towb, P., Bergmann, A., & Wasserman, S. A. (2001). *The protein kinase Pelle mediates feedback regulation in the Drosophila Toll signaling pathway*. Development (Cambridge, England), *128*(23), 4729–4736. Retrieved from, http://www.ncbi.nlm.nih.gov/pubmed/11731453.

Towb, P., Sun, H., & Wasserman, S. A. (2009). Tube Is an IRAK-4 homolog in a Toll pathway adapted for development and immunity. *Journal of Innate Immunity*, *1*(4), 309–321. http://dx.doi.org/10.1159/000200773.

Tsichritzis, T., Gaentzsch, P. C., Kosmidis, S., Brown, A. E., Skoulakis, E. M., Ligoxygakis, P., et al. (2007). A Drosophila ortholog of the human cylindromatosis tumor suppressor gene regulates triglyceride content and antibacterial defense. *Development (Cambridge, England)*, *134*(14), 2605–2614. http://dx.doi.org/10.1242/dev.02859.

Vidal, S., Khush, R. S., Leulier, F., Tzou, P., Nakamura, M., & Lemaitre, B. (2001). Mutations in the Drosophila dTAK1 gene reveal a conserved function for MAPKKKs in the control of rel/NF-kappaB-dependent innate immune responses. *Genes & Development,* *15*(15), 1900–1912. http://dx.doi.org/10.1101/gad.203301.

Vodovar, N., Vinals, M., Liehl, P., Basset, A., Degrouard, J., Spellman, P., et al. (2005). Drosophila host defense after oral infection by an entomopathogenic Pseudomonas species. *Proceedings of the National Academy of Sciences of the United States of America, 102*(32), 11414–11419. http://dx.doi.org/10.1073/pnas.0502240102.

Von Bernuth, H., Picard, C., Jin, Z., Pankla, R., Xiao, H., Ku, C.-L., et al. (2008). Pyogenic bacterial infections in humans with MyD88 deficiency. *Science (New York, N.Y.), 321*(5889), 691–696. http://dx.doi.org/10.1126/science.1158298.

Wang, L., Gilbert, R. J. C., Atilano, M. L., Filipe, S. R., Gay, N. J., & Ligoxygakis, P. (2008). Peptidoglycan recognition protein-SD provides versatility of receptor formation in Drosophila immunity. *Proceedings of the National Academy of Sciences of the United States of America, 105*(33), 11881–11886. http://dx.doi.org/10.1073/pnas.0710092105.

Wang, L., Weber, A. N. R., Atilano, M. L., Filipe, S. R., Gay, N. J., & Ligoxygakis, P. (2006). Sensing of Gram-positive bacteria in Drosophila: GNBP1 is needed to process and present peptidoglycan to PGRP-SA. *The EMBO Journal, 25*(20), 5005–5014. http://dx.doi.org/10.1038/sj.emboj.7601363.

Wang, Z., Wilhelmsson, C., Hyrsl, P., Loof, T. G., Dobes, P., Klupp, M., et al. (2010). Pathogen entrapment by transglutaminase—A conserved early innate immune mechanism. *PLoS Pathogens, 6*(2), e1000763. http://dx.doi.org/10.1371/journal.ppat.1000763.

Waterhouse, R. M., Kriventseva, E. V., Meister, S., Xi, Z., Alvarez, K. S., Bartholomay, L. C., et al. (2007). Evolutionary dynamics of immune-related genes and pathways in disease-vector mosquitoes. *Science (New York, N.Y.), 316*(5832), 1738–1743. http://dx.doi.org/10.1126/science.1139862.

Watson, F. L., Püttmann-Holgado, R., Thomas, F., Lamar, D. L., Hughes, M., Kondo, M., et al. (2005). Extensive diversity of Ig-superfamily proteins in the immune system of insects. *Science (New York, N.Y.), 309*(5742), 1874–1878. http://dx.doi.org/10.1126/science.1116887.

Weber, A. N. R., Gangloff, M., Moncrieffe, M. C., Hyvert, Y., Imler, J.-L., & Gay, N. J. (2007). Role of the Spatzle Pro-domain in the generation of an active toll receptor ligand. *The Journal of Biological Chemistry, 282*(18), 13522–13531. http://dx.doi.org/10.1074/jbc.M700068200.

Weber, A. N. R., Tauszig-Delamasure, S., Hoffmann, J. A., Lelièvre, E., Gascan, H., Ray, K. P., et al. (2003). Binding of the Drosophila cytokine Spätzle to Toll is direct and establishes signaling. *Nature Immunology, 4*(8), 794–800. http://dx.doi.org/10.1038/ni955.

Werner, T., Liu, G., Kang, D., Ekengren, S., Steiner, H., & Hultmark, D. (2000). A family of peptidoglycan recognition proteins in the fruit fly Drosophila melanogaster. *Proceedings of the National Academy of Sciences of the United States of America, 97*(25), 13772–13777. http://dx.doi.org/10.1073/pnas.97.25.13772.

Zhang, S.-Y., Jouanguy, E., Ugolini, S., Smahi, A., Elain, G., Romero, P., et al. (2007). TLR3 deficiency in patients with herpes simplex encephalitis. *Science (New York, N.Y.), 317*(5844), 1522–1527. http://dx.doi.org/10.1126/science.1139522.

CHAPTER THREE

Recent Advances in Septum Biogenesis in *Neurospora crassa*

Rosa Reyna Mouriño-Pérez*, Meritxell Riquelme*,1

Departamento de Microbiología, Centro de Investigación Científica y de Educación Superior de Ensenada, CICESE, Ensenada, Baja California, Mexico
*Both authors contributed equally.
[1]Corresponding author: e-mail address: riquelme@cicese.mx

Contents

1. Introduction	100
2. Septum Ontogeny	101
2.1 Septation during hyphal development	101
2.2 Septation establishment: Regulators and positional markers	103
2.3 Cytoskeletal machinery participating in septum formation	107
3. Septum Wall Biosynthesis	113
3.1 Composition and structure	113
3.2 Septum wall-synthesizing enzymes	113
3.3 Secretory pathway and exocytosis	116
3.4 How is transport of septum-directed vesicles regulated?	118
4. Septal Pores and Associated Proteins	119
5. Conclusions	120
6. Outlook and Future Research	122
Acknowledgments	123
References	123

Abstract

Hyphae of the Ascomycota are tubular cells compartmentalized by perforated septa, whose central pore allows the flow of organelles and cytoplasm. While in plants and yeast septation leads to cell separation, in filamentous fungi the formation of crosswalls appears to have an architectural role, limits the extent of mechanical damage thus maintaining hyphal integrity, and also is of fundamental importance as part of cell differentiation. The increasing number of available fungal genome sequences, knockout mutants, versatile tools for protein tagging, and the continuous improvement of fluorescence microscopes have allowed scientists to analyze living cells and reveal the molecular and cellular basis of septation with unprecedented detail. This review summarizes the recent advances in septum ontogenesis in *Neurospora crassa*. A "septal actomyosin tangle" is the first indication of impending septation. It assembles prior to any visible evidence of plasma membrane inward growth, which occurs concomitantly with the formation and constriction of a contractile actomyosin ring and synthesis of the septum wall. One of the key questions in septum biogenesis is how the septation machinery is assembled to construct a centripetally growing crosswall. Most of the machinery utilized in apical cell wall growth can be expected at septation sites to ensure an organized arrival and supply of vesicles leading to the formation of a septum. Yet, the intrinsically different architecture of the septum may require a different organization and regulation of the wall-synthesizing machinery.

ABBREVIATIONS

CAR contractile actomyosin ring
CHS chitin synthase
SAT septal actomyosin tangle
SIN/MEN septum initiation/mitosis exit network
WB Woronin body

1. INTRODUCTION

Filamentous fungi of the Ascomycota produce elongated tubular cells or hyphae, composed of many compartments that are structurally delimited by simple septa or crosswalls (Gull, 1978). Septa arise as a rim of materials, which develop centripetally leaving a small central pore that in time may become plugged (Bracker, 1967; Hunsley & Gooday, 1974; Trinci & Collinge, 1973). These pores allow the traffic of cytoplasm and organelles through several compartments and contribute to establish a cytoplasmic continuum.

It has been suggested that septa play a structural role in maintaining the tubular shape of a hypha (Rasmussen & Glass, 2005). However, the tubular shape of a hypha is generated at the apex (Bartnicki-García, 2002; Reinhardt, 1892), and while septa may give structural support by adding rigidity to hyphae, most aseptate mutants produce regular hyphal shapes (Gull, 1978). Septa have been shown to have a key role in maintaining hyphal integrity upon injury. The rapid plugging of septal pores avoids an excess of cytoplasmic leakage, facilitating the sealing of the septa and providing a scaffold for the emergence of a new growing axis (Collinge & Markham, 1985; Hunsley & Gooday, 1974; Jedd & Chua, 2000; Momany, Richardson, Van Sickle, & Jedd, 2002; Tenney et al., 2000). Septation in filamentous ascomycetes is essential for differentiation during specific developmental stages of the asexual (conidiation) and the sexual (protoperithecia formation) cycles (Gull, 1978; Raju, 1992). In higher fungi, the existence of perforated septa gives the potential of colony differentiation with the development of specific cell types. The more complex the septal structure, the more complex the type and degree of differentiation (Gull, 1978). Lower fungi achieve limited differentiation due to the lack of septa or the presence of imperforated septa (Gull, 1978).

While classical and molecular genetics studies have uncovered many of the mechanisms underlying nuclear division (Harris, 1997), the basis of septation is only starting to be understood. The combination of fluorescent protein tagging and high-resolution fluorescent live microscopy has allowed elucidation of the different stages of septum formation in several Ascomycetous fungi.

The publication of the *Neurospora crassa* genome sequence (Borkovich et al., 2004; Galagan et al., 2003), as well as the generation of single-gene deletion mutants (Colot et al., 2006), has had a tremendous impact on the progress of Neurospora Cell Biology, as is reflected in the increasing number of recent publications (Riquelme et al., 2011). This review focuses on the current understanding of septum biogenesis in *N. crassa*, based primarily on the data obtained by microscopy analysis of several proteins tagged with fluorescent proteins and analysis of septation mutants. While the information on proteins that participate in septum formation has considerably expanded in the recent years, there is practically no information on how proteins travel and get incorporated into the forming septum.

2. SEPTUM ONTOGENY

2.1. Septation during hyphal development

How the fungal cell decides where and when to form a septum and whether a correlation exists between position of the septum and distance to the apex or between nuclear number and septation has been the foundation for some of the early seminal studies on this process (Gull, 1978; Trinci, 1979).

Cytokinesis (from the Greek: cyto—cell; kinesis—motion, movement) implies the division of one cell cytoplasm to form two daughter cells, a process that begins during the last stages of mitosis in most eukaryotes. Most of the cytokinesis machinery described in yeast cells is found at hyphal septation sites. What is fundamentally different in filamentous fungi is that the final cell separation mechanism is suppressed. Therefore, the term duplication cycle was proposed to describe the collection of events that regulate hyphal growth, mitosis, and compartmentalization (Fiddy & Trinci, 1976). In yeast cells, the primary septum is hydrolyzed by septum-degrading enzymes (SDEs), such as endochitinase 1 (Cts1p) and glucanase (Dse4p)

(Wang, Raniga, Lane, Lu, & Liu, 2009). A similar process occurs during interconidial septation, which involves the formation of a primary and a secondary septum, followed by cell separation. In contrast, in *Candida albicans*, it has been proposed that a hypha-specific cyclin, Hgc1, phosphorylates transcription factor Efg1, which downregulates expression of SDEs (Wang et al., 2009).

Hyphae of filamentous ascomycetes contain several nuclei per compartment, found usually at different mitotic stages. In *Aspergillus nidulans*, nuclei display a parasynchronous mitotic pattern (Clutterbuck, 1970; Rosenberg & Kessel, 1967), and septum formation is spatially controlled by nuclear position and the cell cycle (Harris, 1997, 2001; Momany & Hamer, 1997b; Momany & Taylor, 2000; Wolkow, Harris, & Hamer, 1996). In contrast, in *N. crassa*, asynchronous nuclear divisions prevail (Gladfelter, 2006; Minke, Lee, & Plamann, 1999; Plamann, Minke, Tinsley, & Bruno, 1994), making it difficult to establish a correlation between septum formation and cell cycle.

A collection of *A. nidulans'* temperature-sensitive mutants affected in nuclear distribution, nuclear division, and septation was used to study whether a relationship between these processes determine hyphal compartmentalization (Wolkow et al., 1996). It was shown that the first septum is formed when germlings reach a certain size. The first two rounds of mitosis occur without septum formation. *A. nidulans'* germlings in the preseptation stage can undergo up to a 10-fold increase in size and a eightfold increase in number of nuclei before they form the first septum (Harris, 1997; Harris, Morrell, & Hamer, 1994; Momany & Taylor, 2000). When the germ tube attains a critical size, a nuclear division is sufficient to induce formation of the first septum at the basal end of the germ tube, a process which generally occurs after the fourth mitotic division (Harris et al., 1994). After the first septum is formed, a wave of nuclear division occurs followed by septation (Fiddy & Trinci, 1976). Subsequently, the apical compartment in postseptation cells doubles in size and nuclear number prior to another septation event (Clutterbuck, 1970), and thereafter septa are formed at regular distances after each round of mitosis. In mature hyphae, the formation of the septum closest to the apex occurs at highly variable distances, but the subapical compartments have a uniform length (38 μm on average) (Trinci & Morris, 1979).

Recent analyses have been conducted in growing germlings of *N. crassa* using the vital dye solophenyl flavine as marker of the cell wall to establish when the first and second septa are formed (Y. Valenzuela & M. Riquelme, personal communication). It was found that the length between the germling apex and the most apical septum is highly variable (40–100 μm; $n = 100$), as previously shown (Hunsley & Gooday, 1974). The interseptal distance was more constant, with average values of 48 ± 22 μm ($n = 100$), similar to the previously reported value of 55 μm (Hunsley & Gooday, 1974). In mature hyphae, the distance between septa is 75–95 μm ($n = 40$), and the most apical septum is developed at 180 ± 18 μm ($n = 27$) from the tip (Delgado-Alvarez, Seiler, Bartnicki-Garcia, & Mourino-Perez, 2013). In addition, in both *A. nidulans* and *N. crassa*, a correlation between septation and branch initiation was also observed (Dynesen & Nielsen, 2003; Fiddy & Trinci, 1976; Steele & Trinci, 1977; Trinci, 1979). However, there is practically no understanding on how these two processes are linked.

Septation is an important process for hyphal growth and mycelium differentiation. While aseptate hyphae are viable (see below) and some of the mutants defective in septation show normal hyphal morphology, others display morphological defects, a decrease in vegetative growth, cytoplasmic leakage, or become arrested at early developmental stages (Harris et al., 1994; Kaminsky & Hamer, 1998; Morris, 1975).

2.2. Septation establishment: Regulators and positional markers

2.2.1 The septation initiation network: SIN

As mentioned above, the first steps in the regulatory cascade of septum development seem to be associated with the cell cycle. The molecular machinery underlying septum formation in *N. crassa* and *A. nidulans* is presently being elucidated. Cell cycle-dependent signals of a subset of competent mitotic nuclei activate a signaling cascade known as the septation initiation network (SIN), which is thought to coordinate mitosis and septum formation (Kim, Lu, Shao, Chin, & Liu, 2006; Kim et al., 2009; Liu & Morris, 2000; Momany & Hamer, 1997b; Seiler & Justa-Schuch, 2010; Si, Justa-Schuch, Seiler, & Harris, 2010; Wolkow et al., 1996). This network is equivalent to the mitotic exit network (MEN) described in budding yeast (Bardin & Amon, 2001; Krapp & Simanis, 2005; McCollum & Gould, 2001). The components and architecture of the SIN and the MEN are highly conserved (Grallert, Connolly, Smith, Simanis, & Hagan, 2012). SIN function is essential for septum formation and conidiation during vegetative growth and asexual reproduction (Kim et al., 2006). Deletion of any of the SIN components results in aseptate strains in *A. nidulans* and *N. crassa*

(Liu & Morris, 2000). In *A. nidulans*, SIN components are not required for mycelial growth and colony formation. The deletion of the SIN components *sepH* and *sidB* in *A. nidulans*, orthologues of *S. pombe cdc7* and *sid2*, respectively, produces mycelia with aseptate hyphae without changes in the colonial phenotype (Bruno, Morrell, Hamer, & Staiger, 2001; Kim et al., 2006). This differs from the results in *S. pombe*, where SIN genes are essential (Gould & Simanis, 1997).

The SIN is a network of three types of protein kinases that associate with the spindle pole body through the scaffolding proteins Sid4p, Cdc11p, and Ppc89p (Rosenberg et al., 2006; Tomlin, Morrell, & Gould, 2002). In *N. crassa* activation of SIN is regulated by the RHO-1 GTPase, which in turn is regulated by LRG-1, a specific GTPase-activating protein (GAP). LRG-1 regulates several RHO1 dependent signaling pathways in *N. crassa* including the coordination of polar growth (Vogt & Seiler, 2008).

The Dbf2 subfamily constitutes one of the groups of SIN kinases (Toyn & Johnston, 1994) required for the assembly and constriction of the contractile actomyosin ring (CAR) in future septum sites (Meitinger, Palani, & Pereira, 2012). A second group of kinases, the family of the *n*uclear *D*bf2-*r*elated (NDR) kinases, are key components to induce polarized growth and differentiation in different organisms and require interaction with MOB adapter proteins. MOB–NDR is part of a morphogenesis–related NDR network (MOR) (Gupta & McCollum, 2011; Maerz et al., 2009; Maerz & Seiler, 2010), homologue of the RAM network (regulation of Ace2p and morphogenesis) in *S. cerevisiae* (Nelson et al., 2003). MOR regulates the expression of cell wall–encoding genes, including genes required for efficient cell wall hydrolysis (Bidlingmaier, Weiss, Seidel, Drubin, & Snyder, 2001; Mazanka et al., 2008), and plays a key role in the organization of the actin cytoskeleton (Das, Wiley, Chen, Shah, & Verde, 2009; Das et al., 2007; Schneper, Krauss, Miyamoto, Fang, & Broach, 2004; Vogt & Seiler, 2008). Recently, MST-1, belonging to a third group of kinases, the germinal center (GC) kinases, was characterized in *N. crassa*. MST-1 is thought to function as a SIN-associated kinase that acts in parallel to the central kinase cascade. Additional evidence suggested that MST-1 coordinates SIN and MOR pathways by activating two NDR kinases, COT-1 and DBF-2 (Y. Heilig & S. Seiler personal communication).

Members of the NDR kinases are negative regulators of septation in *N. crassa*. Mutations affecting the NDR kinase COT-1 (Yarden, Plamann, Ebbole, & Yanofsky, 1992), the GC kinase POD-6, regulator of COT-1 (Maerz et al., 2008; Seiler, Vogt, Ziv, Gorovits, & Yarden, 2006), and two MOB-2 proteins result in hyphae with an increased number of septa (Seiler & Plamann, 2003). COT-1 and POD-6 localize to forming septa (Dettmann et al., 2012; Richthammer et al., 2012; Vogt & Seiler, 2008).

2.2.2 Septation landmarks: Rho GTPases and GEFs

Rho (Ras homologous) GTPases are members of the large superfamily of Ras GTPases and function as molecular switches, cycling between active and inactive states (Wennerberg, Rossman, & Der, 2005). Rho GTPases regulate actin organization, cell cycle progression, and gene expression. The formation of the septation apparatus in fungi is at least partially controlled by small Rho-type GTPases, which are activated by specific guanine nucleotide exchange factors (GEFs) (Dunkler & Wendland, 2007; Nakano, Mutoh, Arai, & Mabuchi, 2003; Rasmussen & Glass, 2005; Rossman, Der, & Sondek, 2005; Si et al., 2010).

Increasing evidence shows that the Rho GTPase RHO-4, the GEF BUD-3, and the anillin-related protein BUD-4 are three essential proteins for the establishment of the septation site in *N. crassa* (Justa-Schuch, Heilig, Richthammer, & Seiler, 2010; Rasmussen & Glass, 2005, 2007; Seiler & Justa-Schuch, 2010). RHO-4, BUD-3, and BUD-4 localize exclusively to sites of septation, but not to hyphal apices in *N. crassa*, indicating that none of them is involved in polarized growth (Seiler & Justa-Schuch, 2010).

rho-4 was the first gene required for septation identified and cloned in *N. crassa* (Rasmussen & Glass, 2005, 2007; Rasmussen, Morgenstein, Peck, & Glass, 2008). SIN components are critical for the cortical localization of the RHO-4–BUD-3–BUD-4 module, considered to be the controller of the position of septa (Justa-Schuch et al., 2010; Rasmussen & Glass, 2005; Si et al., 2010).

Although, in general, the cortical markers described in budding yeast are poorly conserved or absent in filamentous fungi (Harris & Momany, 2004), components of the *N. crassa* BUD-3–BUD-4 complex are the closest homologues of the *S. cerevisiae* bud site markers. Prior to the formation of a visible septum as detected by FM4-64 staining (a common membrane marker), but after the appearance of the actomyosin network (septal actomyosin tangle (SAT), described in Section 2.3), both proteins appear as cortical rings at incipient septation sites (Calvert et al., 2011; Delgado-Alvarez et al., 2010; Justa-Schuch et al., 2010; Si et al., 2010). BUD-4 is the first landmark protein to appear at the future septation site as motile cortical dots that subsequently coalesce into a ring. The arrival of BUD-4 coincides with the formation and positioning of the CAR; without BUD-4, the constriction of the CAR cannot progress, as shown by the lack of septa in *bud-4* mutants (Justa-Schuch et al., 2010). BUD-4 has a GTP-binding motif, which as in other anillin-related proteins may assist the BUD-3–RHO-4 complex by supplying it with high levels of GTP (Justa-Schuch et al., 2010; Sanders & Herskowitz, 1996).

In *Ashbya gossypii*, AgBud3 is the first landmark protein described as a determinant of future septation sites (Wendland, 2003). AgBud3 forms a cortical ring preceding septation in the inner edge of the developing septum. In *N. crassa*, localization of BUD-3 depends on the presence of BUD-4 (Justa-Schuch et al., 2010). As in *bud-4* deletion mutants, the lack of *bud-3* results in the absence of septa and hyphal abnormalities, subsequently leading to cytoplasmic leakage and the inability for asexual and sexual reproduction (Justa–Schuch et al., 2010).

In *A. nidulans*, AnBud4 seems to have a different function. It is present in septation sites in both hyphae and conidiophores; however, it is not essential for septum formation. Septum formation is only delayed in *bud4* mutants (Si et al., 2010). In contrast, *bud3* mutants are viable but aseptated (Si et al., 2010). It seems that contribution of AnBud3 and AnBud4 to septum formation is independent in *A. nidulans*. As in *N. crassa*, AnBud3 acts as the Rho4 GTPase GEF and AnBud4 seems to participate in septin organization (Si et al., 2010), but this remains to be established.

Some of the machinery involved in polarized apical growth is also found at septal disks and septal pores (Lichius, Yanez-Gutierrez, Read, & Castro-Longoria, 2012; Schurg, Brandt, Adis, & Fleissner, 2012; Vogt & Seiler, 2008). BUD-6, a polarisome component, is recruited to the incipient septation site preceding plasma membrane growth observed by FM4-64 (Lichius et al., 2012). BUD-6 remains associated with the leading edge of the CAR and concentrates surrounding the septal pore, persisting for hours. This suggests that BUD-6 might interact with the formin BNI-1 as part of the machinery of CAR anchoring and constriction, also having additional roles at the septal pore after the septum is totally formed (Lichius et al., 2012). The CDC-42–RAC–CDC-24 GTPase module is present in the apex of *N. crassa* and is required for polarized growth and hyphal morphogenesis (Araujo-Palomares, Richthammer, Seiler, & Castro-Longoria, 2011). In addition, both RAC and CDC-42 were observed also at developing septa in *N. crassa* (Araujo-Palomares et al., 2011). Cdc42p and its homologues seem to have a consistent function during septation and cell division in yeasts (Park & Bi, 2007; Perez & Rincon, 2010), in the filamentous ascomycetes *Penicillium marneffei* (Boyce, Hynes, & Andrianopoulos, 2005), *A. niger*

(Kwon et al., 2011), and the dimorphic ascomycete *C. albicans* (Bassilana, Hopkins, & Arkowitz, 2005). Although the specific function of RAC and CDC-42 in septum development in filamentous fungi remains to be clarified, it has been suggested that the components of the CDC-42–RAC–CDC-24 module are negative regulators of the septation process (Araujo-Palomares et al., 2011) and seem to have an opposed function to that of RHO-1 and RHO-4 that positively contribute to septation (Justa-Schuch et al., 2010; Rasmussen & Glass, 2005; Seiler & Justa-Schuch, 2010; Vogt & Seiler, 2008).

Cell morphogenesis in *N. crassa* is influenced through the phospholipid-dependent organization at the apical plasma membrane. The phosphatidylinositol 4,5-bisphosphate (PtdIns(4,5)P$_2$ or PIP$_2$) forms plasma membrane microdomains that support polar growth. The enzyme responsible for this accumulation is the PI4P 5-kinase (MSS-4). In *N. crassa*, MSS-4 localizes in the plasma membrane in sites of active growth. At the hyphal apex, MSS-4 adopts a ring-like distribution immediately behind the apex. In septa, it localizes to sites of septum initiation, decorating the inner edges of the constricting ring of the cell wall (Mahs et al., 2012).

2.3. Cytoskeletal machinery participating in septum formation
2.3.1 Actin and actin-binding proteins: SAT, CAR, and actin patches
The actin cytoskeleton has a key role during septum formation in different filamentous fungi (Berepiki, Lichius, Shoji, Tilsner, & Read, 2010; Delgado-Alvarez et al., 2010; Girbardt, 1979; Hoch & Howard, 1980; Roberson, 1992). Unlike budding yeast, where actin relocalizes from bud tips to bud necks during cytokinesis, in filamentous fungi it localizes simultaneously to hyphal tips and septa (Harris et al., 1994). Actin seems to be the first protein to localize at future septation sites (Berepiki et al., 2010; Delgado-Alvarez et al., 2010; Momany & Hamer, 1997b). High concentrations of actin have been found at the sites where septa will form (Butt & Heath, 1988; Calvert et al., 2011; Runeberg, Raudaskoski, & Virtanen, 1986; Salo, Niini, Virtanen, & Raudaskoski, 1989; Tanabe & Kamada, 1994). Actin appears first as a ring; next, it forms an invagination band with cell wall deposition in the furrow, in a process that depends upon microtubule integrity. The use of actin inhibitors such as Cytochalasin A blocks septum formation. This effect can be reversed by washing out the inhibitor, which suggests that continued actin polymerization is required for septation (Harris et al., 1994). Detailed time course experiments have shown that before there is any sign of septum formation, namely, the centripetal growth of the plasma membrane toward the hyphal center, actin made structures

appear at the septation sites (Berepiki et al., 2010; Delgado-Alvarez et al., 2010; Momany & Hamer, 1997b).

Actin and actin-binding proteins (ABPs) arrange into diverse higher-order structures, which participate in different stages of septum development. Three different actin assemblies have been recognized during septal development in *N. crassa* (Mouriño-Pérez, 2013): a loose network of actin cables (Septal actomyosin tangle or SAT), a CAR and two sets of actin patches flanking the septum. All of them are transiently assembled, appearing at different stages of septal development (Berepiki et al., 2010; Delgado-Alvarez et al., 2010).

In *N. crassa* by using Lifeact-GFP as an actin reporter, a broad interlace of F-actin cables was observed before any evidence of plasma membrane ingrowth (Berepiki et al., 2010; Delgado-Alvarez et al., 2010). This large tangle of actin has been named the SAT (Mouriño-Pérez, 2013) (Figure 3.1A). After 2 or 3 min, the SAT gradually coalesces to form a single ring, the CAR. Septal membrane growth coincides with the constriction of the CAR. While most membrane markers are found distributed through the growing septal ring, CAR moves from the periphery to the center of the ring and disappears, leaving only a few patches (Delgado-Alvarez et al., 2013; Figure 3.1B–D).

ABPs, such as myosin, formin, and tropomyosin, delimit the structure of the actin microfilaments and provide the machinery needed for constriction. Myosins are motor proteins associated with actin, involved in contraction and in a wide range of motility processes. The myosin proteins MYO-2 (Class II) and CDC-4 (MYO-2-light chain) provide the contractile force for SAT coalescence and CAR constriction. They seem to align with the Lifeact-GFP-labeled cables and also colocalize with some actin patches (Delgado-Alvarez et al., 2013). During CAR constriction, MYO-2 occupies the center of the septal plate (Calvert et al., 2011). In *A. nidulans*, MyoB (Class II) is required for septation, but not MyoE (Class V) that is responsible for vesicle transport into the Spitzenkörper (Taheri-Talesh, Xiong, & Oakley, 2012). Tropomyosin (TPM-1) is a protein that regulates the interaction between actin filaments and myosin. By tagging it with GFP, TPM-1-GFP was shown to be associated with SAT during its formation and remains as part of CAR as long as actin is present (Figure 3.1E–G).

A third actin structure comprises the patches associated with the ARP-2/3 complex, fimbrin, and coronin. These proteins are found in subapical endocytic patches (Delgado-Alvarez et al., 2010; Echauri-Espinosa, Callejas-Negrete, Roberson, Bartnicki-Garcia, & Mourino-Perez, 2012;

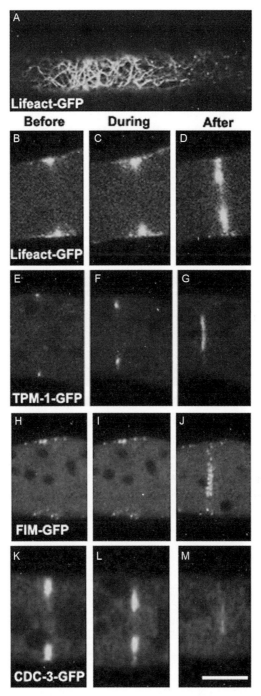

Figure 3.1 Comparative localization of actin cytoskeleton components during septum development in mature hyphae of *N. crassa* imaged by laser scanning confocal microscopy. (A) Actin visualized with the reporter Lifeact-GFP showing the septal actomyosin

(Continued)

Upadhyay & Shaw, 2008). Actin, together with the ARP-2/3 complex, fimbrin, and coronin, forms scattered patches that accumulate in the cortical cytoplasm at the site of septum development before membrane growth (Delgado-Alvarez et al., 2010). A cortical double ring of patches forms afterward at each side of the septum and disappears once septation is completed (Delgado-Alvarez et al., 2010) (Figure 3.1H–J). The patches participate in late stages of septation but do not participate in the assembly of the CAR. These patches seem to be part of the endocytic machinery (Upadhyay & Shaw, 2008) and may contribute to membrane remodeling in the septum during its formation.

The "search, capture, pull, and release" model in *S. pombe* is similar to the conversion process from SAT to CAR in *N. crassa* (Kovar, Sirotkin, & Lord, 2011; Lord, 2010). In *S. pombe*, cytokinesis starts with the accumulation of the anillin-like Mid1 nodes in the division site (Bahler et al., 1998; Chang, Woollard, & Nurse, 1996; Sohrmann, Fankhauser, Brodbeck, & Simanis, 1996). Close to the beginning of mitosis, Mid1 recruits Myo2, Cdc4, and Rlc1 (Pollard, 2008; Vavylonis, Wu, Hao, O'Shaughnessy, & Pollard, 2008; Wu, Kuhn, Kovar, & Pollard, 2003; Wu et al., 2006) and also IQGAP Rng2, F-BAR protein (Cdc15), formin (Cdc12), tropomyosin (Cdc8), and α-actinin (Ain1) (Coffman, Nile, Lee, Liu, & Wu, 2009; Wu et al., 2003, 2006). The formin associated with the Mid1 nodes nucleates actin filaments that associate with tropomyosin and rapidly elongates. The actin filaments become cross-linked into short antiparallel bundles that finally coalesce to form the CAR (Kamasaki, Osumi, & Mabuchi, 2007). While in *A. nidulans* microtubules are required for the initiation and progression of septation (Momany & Hamer, 1997b), *N. crassa* can form septa when microtubules are depolymerized (Sánchez-León et al., 2011).

Figure 3.1—Cont'd tangle (SAT) in the cortical region of the hypha previous to the plasma membrane growth. A meshwork of thick actin cables can be observed before they coalesce into a contractile actomyosin ring (CAR). (B–M) Distribution of different actin cytoskeleton markers and a septin before (first column), during (second column), and after CAR formation (third column). (B–C) Lifeact-GFP is broadly distributed in the septal plate. (E–G) Tropomyosin labeled with GFP (TPM-1-GFP) is present as part of the CAR, occupying the inner edge of the ring. (H–J) Fimbrin-GFP (FIM-GFP) is not part of the CAR; it is part of the actin patches presumably associated with endocytosis and it appears a few seconds before CAR formation and forms a double ring flanking the septal plasma membrane. (K–M) Septin CDC-3 tagged with GFP localizes at the CAR from the moment it starts to be formed and disappears soon after septum completion. Scale bar = 10 μm. (For color version of this figure, the reader is referred to the online version of this chapter.)

2.3.2 Actin polymerization proteins: Formins

Formins are actin nucleating proteins characterized by the presence of a highly conserved FH2 domain, which is necessary for actin filament assembly *in vitro* and *in vivo* (Evangelista, Pruyne, Amberg, Boone, & Bretscher, 2002; Kovar, 2006; Pruyne et al., 2002; Sagot, Rodal, Moseley, Goode, & Pellman, 2002). The FH2 domain nucleates actin filaments and enables their polymerization (Harris, Li, & Higgs, 2004; Kovar, 2006; Moseley et al., 2004; Zigmond et al., 2003). Formins are attached to the plus ends of actin filaments and can generate linear arrays such as actin cables or the CAR (Higashida et al., 2004; Kovar, 2006; Kovar & Pollard, 2004; Pruyne et al., 2002; Romero et al., 2004). Formins also have a proline-rich FH1 domain, which binds profilin (Chang et al., 1996), a protein that facilitates actin subunit delivery to the growing ends of actin filaments capped by formins (Kovar, 2006; Sagot et al., 2002).

Null or conditional mutants for formins BNI-1 and SepA result in aseptate phenotypes in *N. crassa* and *A. nidulans*, respectively (Harris, Hamer, Sharpless, & Hamer, 1997; Justa-Schuch et al., 2010; Lichius et al., 2012; Sharpless & Harris, 2002). These formins, as in budding yeast, are essential for actin ring assembly (Tolliday, VerPlank, & Li, 2002; Yoshida et al., 2006). BNI-1 is present both at the hyphal apex and in developing septa (Lichius et al., 2012). It is recruited to the incipient septation site before membrane ingrowth became visible by FM4-64 staining but after SAT formation and concomitant with the establishment of the CAR. This suggests that the polymerization of the actin cables forming the SAT at septation sites is either formin-independent or occurs in other regions of the hyphae (Huang et al., 2012).

2.3.3 Septins

Septins were first discovered in *S. cerevisiae* (Hartwell, 1971; Longtine et al., 1996). They constitute a conserved family of GTP-binding proteins found in eukaryotes from yeasts to mammals. They are able to polymerize and form hetero-oligomers that assemble into complex structures, whose detailed molecular architecture has been described in different organisms (Field & Kellogg, 1999). The different septin monomers assemble into oligomers, which in turn form higher-order structures called multiseptin complexes. These structures are composed of different numbers of subunits, depending on the organism. In filamentous fungi, septins function as platforms recruiting and organizing other proteins at sites of polarized growth and septum formation (DeMay, Meseroll, Occhipinti, & Gladfelter, 2009; Harris, 2001; Lindsey, Cowden, Hernandez-Rodriguez, & Momany, 2010; Momany,

Zhao, Lindsey, & Westfall, 2001, Berepiki & Read, 2013) and are currently viewed as cytoskeletal elements (Versele & Thorner, 2005). *N. crassa* has six predicted septins (ASP-1 [NCU01998], ASP-2 [NCU06414], CDC-3 [NCU08297], CDC-10 [NCU03515], CDC-11 [NCU02464] and CDC-12 [NCU03795]). Until recently, only the localization of CDC-3 at the hyphal subapical collar and at forming septa, had been reported (Fig. K-M; Sánchez, Freitag, Smith, & Riquelme, 2009). Newly, CDC-10, -11, -12, and ASP-1 have been also GFP tagged and localized during different developmental stages in *N. crassa* and were shown to form rings, fibers, bar-like structures and caps (Berepiki & Read, 2013). Except for *asp-1* and *asp-2*, the other septin deletion mutants showed altered morphologies and abnormal germ tube emergence.

A. *nidulans* has five septins (AspA, AspB, AspC, AspD, and AspE). All of them are expressed during vegetative growth. AspB is the septin expressed at higher levels (Momany et al., 2001). It forms a single ring and bars overlapping the CAR at the septation site (Momany & Hamer, 1997a; Westfall & Momany, 2002). AspB is nonessential but is required for normal growth and morphogenesis in *A. nidulans*. The *aspB* mutant forms septa although more slowly than the wild type (Hernandez-Rodriguez, Hastings, & Momany, 2012). AspB in *A. nidulans* and CDC-3 in *N. crassa* are present as rings at septation sites after SAT formation (Momany & Hamer, 1997a; Figure 3.1K–M). AspB does not disassemble immediately after septum assembly; it remains in septa, but only on the side facing the leading compartment (Momany & Hamer, 1997a). This behavior is similar to that observed by Sep7 in *C. albicans* (Gonzalez-Novo et al., 2008). It has been speculated that septins function as repressors or inhibitory markers for cell separation. However, this needs to be further investigated. Deletion mutants of *aspA* and *aspC* showed that both are involved in septation, but are not essential. *aspA* mutants display delayed or reduced septation at the beginning of incubation, but after a few hours can form normal septa. However, *aspC* mutants make only one third of the number of septa of the wild type and these septa appeared to be abnormal (Lindsey et al., 2010). There are identical localization patterns of AspA-GFP and AspC-GFP forming rings or collars in septa. These septins may have similar functions. Although AspA seems to be a central septin needed for the retention of AspC and AspB in the *A. nidulans* septin complex, AspC does not localize in septa in *aspA* mutants, and also all AspB rings, bars, and filaments are lost in the absence of AspA (Hernandez-Rodriguez et al., 2012). These differences in functions between AspA and AspC are also shown in the increased severity of all phenotypes in the *aspA/aspC* double mutant (Lindsey et al., 2010).

3. SEPTUM WALL BIOSYNTHESIS

3.1. Composition and structure

Most of the electron micrographs showing sections of fungal hyphae at the septal region come from works published in the 1960s and 1970s. The seminal work of Hunsley and Gooday (1974) gives insight into fine structure and chemical composition of septa in *N. crassa*.

The cell wall of *N. crassa* is primarily composed of two layers: an outer layer of glucan–peptide–galactosamine and an inner layer made up of β-1,3-glucan with an imbedded internal core of chitin fibrils (Free, 2013; Mahadevan & Tatum, 1967). In the septum region, the glucan–peptide layer is thicker, although it does not enter significantly the crosswalls (Hunsley & Gooday, 1974; Mahadevan & Tatum, 1967). Electron microscopy of *N. crassa* septa showed mainly microfibrillar material arranged primarily in a tangential orientation (Hunsley & Gooday, 1974). Autoradiography analysis and chemical and enzymatic extractions suggested that chitin is a major component of the septum (Hunsley & Gooday, 1974). Light microscopy autoradiography indicated the incorporation of not only *N*-acetylglucosamine, the chitin constituent, but also glucose, the β-1,3-glucan constituent, into developing septa at a higher density than the surrounding lateral walls (Hunsley & Gooday, 1974). However, the relative grain density was about three times higher for *N*-acetylglucosamine than for glucose. In addition, surrounding the lateral cell wall in septal regions, proliferation of a reticulum thought to be glycoprotein was observed in 24-h cultures (Hunsley & Gooday, 1974).

3.2. Septum wall-synthesizing enzymes

Chitin synthesis is catalyzed by enzymes belonging to the family of chitin synthases (CHSs), which incorporate subunits of *N*-acetylglucosamine (GlcNAc) into linear chains of β (1,4)-GlcNAc (Figure 3.2). Filamentous ascomycetes have single encoding genes for each of the seven reported classes of CHS (Borkovich et al., 2004; Choquer, Boccara, Goncalves, Soulie, & Vidal-Cros, 2004; Riquelme & Bartnicki-García, 2008). CHSs have been localized at sites of growth, including hyphal apices, branches, and forming septa. In *N. crassa*, all CHSs tagged with fluorescent proteins have been found at developing septa in vegetative hyphae (Riquelme et al., 2007; Sánchez-León et al., 2011). Most CHSs accumulated first as a cylindrical sleeve under the existing lateral wall at sites of future septum

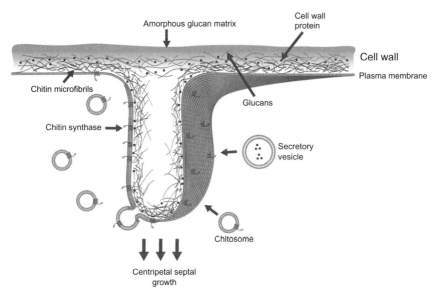

Figure 3.2 Three-dimensional illustration of a septum section of *N. crassa* portraying the main components of the wall and the machinery involved in wall synthesis. (For color version of this figure, the reader is referred to the online version of this chapter.)

formation, and they gradually entered into the centripetally growing septal ring. CHS-2 (Class II) was the only CHS localized at the rim of the developing septum and remained at the septal pore (Fajardo-Somera et al., 2013). Disruption of the actin cytoskeleton by Cytochalasin A or Latrunculin A caused an uneven localization of CHS-1-GFP at septa (Sánchez-León et al., 2011). In coexpression experiments, it was shown that CHS-1-mChFP partially colocalizes with the CAR labeled by Lifeact-GFP during its constriction (Figure 3.3). In *A. nidulans* where some CHSs have been analyzed at the cellular level, ChsA, ChsB, ChsC, CsmA, and CsmB (belonging to CHS classes II, III, I, V, and VI, respectively) have been also found in septa (Fukuda et al., 2009; Horiuchi, 2009; Ichinomiya, Yamada, Yamashita, Ohta, & Horiuchi, 2005; Takeshita, Ohta, & Horiuchi, 2005). ChsA, ChsB, and ChsC associate with CAR as it constricts, whereas CsmA and CsmB, which contain an N-terminal myosin motor-like domain, seem to function in regulation of chitin deposition around the septal pore.

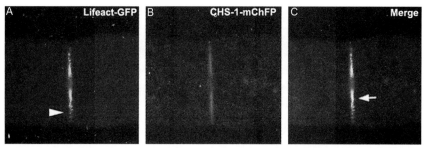

Figure 3.3 Coexpression of the actin reporter Lifeact-GFP and the chitin synthase CHS-1 tagged with mChFP during septum formation in *N. crassa* imaged by laser scanning confocal microscopy. (A) Localization of Lifeact-GFP during the contraction of the CAR; note both the CAR and the actin patches (arrowhead). (B) Localization of CHS-1-mChFP in the septum. (C) Overlay of (A) and (B); arrow indicates the region where the actin ring and the CHS-1 ring colocalize; the inner part of the ring is exclusively made of actin, and also there are actin patches (arrowhead) surrounding the distal part of the septum. Scale bar = 10 μm. (For color version of this figure, the reader is referred to the online version of this chapter.)

The time course for formation of individual septa should vary depending on the hyphal diameter. For *N. crassa* hyphae of 7 μm in diameter, the time was estimated to be 3.5–4 min (Hunsley & Gooday, 1974). By using CHS as markers, septum development lasted 3–6 min (Riquelme et al., 2007; Sánchez-León et al., 2011). Once completed, no more CHS arrived to the septum as proven by FRAP analysis.

β-1,3-Glucan synthesis is catalyzed by the glucan synthase complex, GSC, composed at least of a catalytic subunit (FKS) and a regulatory subunit (RHO GTPase) (Kang & Cabib, 1986; Latgé & Calderone, 2006; Lesage et al., 2004). FKS is a transmembrane protein, and once at the membrane, it accepts the glucose moiety from the cytoplasm and incorporates it into the growing β-1,3-glucan chains (Douglas, 2001). Filamentous fungi have a single FKS encoding gene, whereas *S. cerevisiae* and *S. pombe* have three and four genes, respectively, with homology to FKS (Lesage et al., 2004). In both unicellular ascomycetes, FKS was shown to localize at growing sites and septa. In *S. cerevisiae*, Fks1p and Fks2p are found at septa (Dijkgraaf, Abe, Ohya, & Bussey, 2002). In *S. pombe*, Bgs1/Csp1 and Bgs3go are also found at septa in a SIN-dependent manner (Cortes, Ishiguro, Durán, & Ribas, 2002). In *N. crassa* while components of the GSC have been found at growing apices, none of them have been found at developing septa (Richthammer et al., 2012; Verdín, Bartnicki-Garcia, & Riquelme, 2009).

3.3. Secretory pathway and exocytosis

Most literature describes septum formation as a constriction of the cell coupled with membrane invagination or plasmalemma infolding. Strictly, this is not an accurate description of the biological process accompanying septum formation. While it is true that transmission electron micrographs showed membrane indentation at septal regions, recent studies clearly show the need of new membrane delivery during septum formation (Schiel & Prekeris, 2013), and therefore this process should be envisioned as an inward growth of the membrane that occurs centripetally by targeted fusion of vesicles (Figure 3.2). Two main populations of vesicles differing in size and involved in fungal cell wall growth have been described in fungal hyphae (Bartnicki-García, 1987, 1990). Macrovesicles or secretory vesicles have diameters on average larger than 100 nm, whereas microvesicles are smaller than 100 nm in diameter. CHS and GSC are transported in different populations of vesicles, which at the apex accumulate at distinct layers of the Spitzenkörper (Verdín et al., 2009). GSCs are found in macrovesicles that accumulate in the outer layer of the Spitzenkörper, while CHSs are transported in a specialized population of microvesicles, called chitosomes, that accumulate at the core of the Spitzenkörper and can synthesize chitin microfibrils *in vitro* (Bracker, Ruiz-Herrera, & Bartnicki-Garcia, 1976). Although, as mentioned above, all CHSs are found at the forming septum ring, there is no mention of chitosomes in the vicinity of septa in the literature. However, early transmission electron micrographs showed that developing septa of *N. crassa* were surrounded by vesicles ranging from 40 to 95 nm in diameter, smaller than the macrovesicles found at hyphal apices ranging from 60 to 150 nm (Hunsley & Gooday, 1974), that could correspond to chitosomes. In addition, in electron micrographs of *Fusarium acuminatum*, both microvesicles and macrovesicles could be observed near septa (Howard, 1981). Recently, putative chitosomes adjacent to a septum have been observed in germlings of *A. nidulans* (R. Roberson, personal communication).

Vesicle fusion is a very fast process (occurring in the order of milliseconds), much faster than the best cell ultrastructure-preserving techniques used for sample fixation in electron microscopy, including cryofixation (Chandler & Roberson, 2008). Furthermore, preservation of cytoplasmic ultrastructure in hyphal distal regions is particularly difficult (R. Roberson, personal communication). This could explain the scarce availability of electron micrographs displaying vesicles nearby or fusing with septa.

In addition to cell wall-synthesizing enzymes, other plasma membrane proteins have been localized at septa. In *A. oryzae* and *N. crassa*, the plasma membrane amino acid transporters AoUapC and AoGap1, and the plasma membrane proton pump PMA-1, respectively, were found at septa and the lateral plasma membrane, but not at the Spitzenkörper (Fajardo-Somera, Bowman, & Riquelme, 2013; Hayakawa, Ishikawa, Shoji, Nakano, & Kitamoto, 2011) (Figure 3.4). Unlike the cell wall-synthesizing enzymes, which get incorporated to developing septa, these enzymes were incorporated at already formed septa. This suggests that either (1) the vesicles carrying these enzymes fuse with the plasma membrane of the already existing septa by a mechanism different than full fusion and collapse of vesicle that does not contribute to gain of membrane or (2) the vesicles indeed fuse completely, thereby introducing an excess of membrane that needs to be retrieved by endocytosis.

Secreted proteins, such as the α-amylase AmyB and the glucoamylase Gla, have been found in septa in addition to the tips in *A. oryzae* and *A. niger* (Gordon et al., 2000; Hayakawa et al., 2011; Masai, Maruyama, Nakajima, & Kitamoto, 2003). This opens up a new window of research, since secretion into the periplasm or external medium has been linked

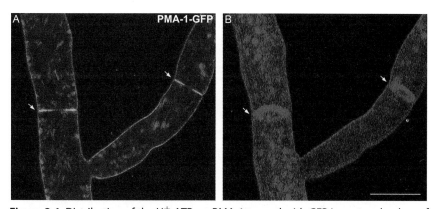

Figure 3.4 Distribution of the H^+-ATPase PMA-1 tagged with GFP in mature hyphae of *N. crassa*. (A) Image corresponding to a single plane of the basal part of a branched hypha expressing PMA-1-GFP and recorded by laser scanning confocal microscopy. H^+-ATPase can be seen through all the plasma membrane, including the already formed septal membrane as shown by arrows. (B) Three-dimensional reconstruction of the hypha in (A) from z-stacks, where it is possible to observe the perforated septal membrane. Scale bar = 10 μm. (For color version of this figure, the reader is referred to the online version of this chapter.)

primarily to growing tips, where the secreted enzymes digest complex compounds into nutrients that can be absorbed by the hyphae and distributed along the colony.

3.4. How is transport of septum-directed vesicles regulated?

Vesicles are basic carriers that traffic from one cellular compartment to another and are targeted to specific cellular locations. The conventional secretory pathway begins with proteins synthesized in the endoplasmic reticulum (ER) that are later sorted through Golgi to their final destination, that is, plasma membrane, organelles, or external medium. The process of vesicle biogenesis and traffic includes (1) vesicle budding (scission from the donor membrane-vesicle formation), (2) vesicle transport, (3) vesicle docking (vesicle and target membrane brought into close proximity), and (4) vesicle fusion (Cai, Reinisch, & Ferro-Novick, 2007; Novick et al., 2006). Vesicular transport and active sorting between organelles occur through a succession of events that require the coordinated action of coating complexes, for vesicle budding and recognition of cargo-sorting signals, tether complexes that interact with coating proteins and mediate docking, and soluble-NSF (N-ethylmaleimide-sensitive-factor) attachment protein receptors (SNAREs) that facilitate the fusion between donor and target membranes (Bonifacino & Glick, 2004). These processes are regulated by Rab GTPases, another subfamily within the super family of Ras GTPases involved primarily in vesicular transport and organelle dynamics, and motor proteins. In yeast and mammals, the traffic events from the ER to Golgi, including the mechanisms that maintain compartment identity and ensure fidelity of transport, have been well characterized (Barrowman, Bhandari, Reinisch, & Ferro-Novick, 2006), while post-Golgi traffic is much less understood.

In fungal hyphae, studies toward dissecting the components of the secretory pathways and elucidating traffic of membranes have emerged only recently (Bowman, Draskovic, Freitag, & Bowman, 2009; Maruyama, Kikuchi, & Kitamoto, 2006; Pantazopoulou & Peñalva, 2009; Shoji, Arioka, & Kitamoto, 2008), and only a few focus on septum-directed secretion (Hayakawa et al., 2011). In fact, no specific Rab GTPase has been identified thus far regulating septum-directed vesicular traffic.

Using the ER chaperone BipA-EGFP as a marker of the ER in *A. oryzae*, accumulation of fluorescence was observed in two parallel lines flanking the septum, and as a tubular network along the septum once it was formed

(Maruyama et al., 2006). Further studies are needed to analyze if the ER localized at septal regions contributes directly to local protein synthesis.

Exocytosis is the last step of the secretory pathway and it involves the fusion of vesicles with the plasma membrane, a process that, in fungi, ensures the delivery of cell wall-synthesizing enzymes, membrane proteins, and lipids in areas of active growth. Exocytosis occurs in all regions where cell wall growth is occurring, including nonapical regions (Read, 2011). All processes involving membrane fusion require the action of tethers to bring closer at the appropriate sites the donor and target membranes. The exocyst is an octameric-tethering factor involved in vesicle docking with the plasma membrane at polarized sites of growth in all eukaryotic cells (TerBush, Maurice, Roth, & Novick, 1996). In the budding yeast, the exocyst has been implicated in a variety of cellular processes including cytokinesis. Post-Golgi vesicles and exocyst components localize at the bud neck just before the spindle disassembly and actomyosin ring contraction (VerPlank & Li, 2005). During cytokinesis, exocyst components concentrate first as a ring and then as two rings at the neck that separates the mother cell and the bud (Hsu, TerBush, Abraham, & Guo, 2004). Within the filamentous fungi, only *A. oryzae* has displayed a component of the exocyst, AoSec3, at septa (Hayakawa et al., 2011).

After vesicle docking, interaction between v-SNAREs (vesicles) and t-SNAREs (target membrane) mediates the final fusion event between the donor and target membranes (Pelham, 1999). In *A. oryzae*, SNAREs have been localized at septa (Hayakawa et al., 2011; Kuratsu et al., 2007). The only v-/t-SNAREs studied in *N. crassa* have been localized at the hyphal apex (Gupta & Heath, 2000). Further studies are required to identify septum-specific SNAREs.

4. SEPTAL PORES AND ASSOCIATED PROTEINS

N. crassa septal pores often are between 350 and 500 nm in diameter, depending on the stage of completion. Little is known about the constituents of the septal pore in filamentous fungi and how the diameter of the pore is regulated. Pores must maintain the smallest diameter that would allow quick plugging in response to injury, while at the same time, when open, permit the traffic of organelles across compartments. A heterogeneous group of proteins named SPA proteins (Septal pore-associated proteins) with diverse localization and functions appear to be involved in pore-rim lining and occlusion. SPA aggregation offers structural plasticity for pore gating due to the variability of pore diameter. Nevertheless, understanding how the SPA proteins contribute to the maintenance of the septal pore is an interesting area for further investigation (Lai et al., 2012).

As mentioned earlier, upon injury of the hypha, the septal pores can be rapidly plugged, a phenomenon that avoids an excess of cytoplasmic leakage, facilitates the sealing of the septa, and allows the rapid regeneration of a new growing axis. Woronin bodies (WBs) are a specialized class of peroxisomal-derived organelles that quickly plug the septal pore upon injury of the hypha, avoiding the excessive loss of cytoplasm (Fleissner et al., 2005; Jedd & Chua, 2000; Liu et al., 2008). WBs are manufactured continuously in differentiated apical peroxisomes. When septation initiates, WBs attach to the cell cortex in subapical compartments. Tip lysis triggers WB release from cortex, septal pore plugging, membrane resealing, and new tip growth (reviewed by Pieuchot & Jedd, 2012). The positioning of WBs in the cortex and at septal pores, when signals from cellular damage induce release, is coordinated by Leashin tethers LAH-1 and LAH-2, respectively (Ng, Liu, Lai, Low, & Jedd, 2009).

A reniform electron-dense structure lining the septal pore different than the septal plug has also been reported (Hunsley & Gooday, 1974) and classified as a second system of pore gating (Jedd & Pieuchot, 2012; Trinci & Collinge, 1973).

A diverse array of septal pore-associated proteins has been identified (Fleissner & Glass, 2007; Maruyama, Escano, & Kitamoto, 2010; Rasmussen & Glass, 2007; Takeshita, Vienken, Rolbetzki, & Fischer, 2007). In a recent study, mass spectrometry analysis of WB-associated proteins allowed the identification of 17 septal pore-associated proteins (Lai et al., 2012). SO (soft; fs-n; ham-1), a protein involved in hyphal fusion (Fleissner, Leeder, Roca, Read, & Glass, 2009; Fleissner et al., 2005), has also a role in septal pore sealing (Fleissner & Glass, 2007). Upon hyphal injury, SO accumulates at septal pores in a WB-dependent manner. It has a WW domain, a two-conserved tryptophan domain that binds to proline-rich peptide motifs. Additionally, as a result of heterokaryon incompatibility (program cell death) and aging in the interior of the colony, SO plugging occurs in a WB-independent manner.

5. CONCLUSIONS

Although the main mechanisms for septum formation seem to be conserved between unicellular organisms and filamentous fungi, this review points out some of the differences in the regulation of septum initiation, the selection of the septation sites, the organization and constriction of CAR, the cell wall construction, and plasma membrane remodeling in the fungus *N. crassa*. Given the recent evidence showing that the same set of synthases operates in both septation sites and apical sites, the challenge is to elucidate the factors responsible for organizing the delivery and operation of these enzymes to produce a different wall architecture. As described here, while many of the proteins involved in polarization and cell wall growth at the apex have been identified participating also in different stages of septation, their key regulators differ (Figure 3.5). For a septum to be formed, two different processes have to occur: first, nuclear division needs

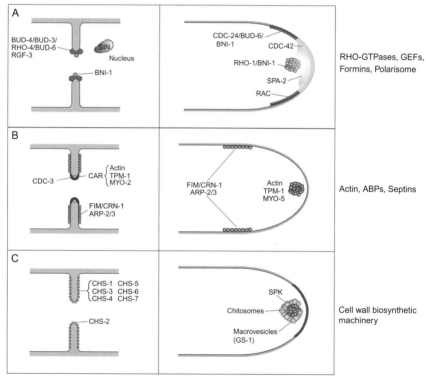

Figure 3.5 Scheme summarizing the main components that orchestrate membrane and wall growth at septa and apices of *N. crassa*. Similarities and differences between the apical and the septation machineries can be observed. In septa, the presence of all components is transient and it is possible to establish a chronology during the finite septal growing process; in contrast, at the apex, the growing process is continuous. Although some components are found at both locations, their regulation differs. (For color version of this figure, the reader is referred to the online version of this chapter.)

to activate the SIN cascade, and second, recruitment of structural and regulatory elements for cell wall construction and plasma membrane growth is required. In *N. crassa*, F-actin is the first protein observed at the future septation site; actin associates with tropomyosin and MYO-2 to form a tangle of thick cables, the SAT. What leads actin to form a SAT at specific sites remains to be identified. Proteins of the SIN components, the BUD-4–BUD-3–RHO-4 complex and the formin BNI-1 (Justa-Schuch et al., 2010; Seiler & Justa-Schuch, 2010), participate in the transition of SAT into a CAR and its posterior constriction.

In summary, the steps to form a septum include (1) accumulation of a broad band of actin surrounding the future septation site (SAT), (2) the addition of contractile force through MYO-2, (3) formation and anchoring of the CAR, (4) CAR constriction, and (5) inward growth of plasma membrane, coupled with construction of the cell wall.

6. OUTLOOK AND FUTURE RESEARCH

There are several open questions on septal development in *N. crassa* that need to be further addressed. These remaining questions can be classified into four categories. The first category involves assessing whether in *N. crassa* there are compulsory links between cell cycle and septum formation. Although in other filamentous fungi it is relatively easy to follow the cell cycle in association with septum formation, in *N. crassa* it is still not clear how the nuclear signals may contribute mechanistically to the regulation of the process *in situ*. It has been proposed that all cortical regions may potentially undergo septum formation. However, the assembly of a septal band at one specific site might block the ability of adjacent mitotic nuclei to trigger the same event.

The second category relates to the regulation of the actin cytoskeleton. It is important to identify what induces SAT formation and why proteins such as the formin BNI-1 are observed after actin cables are polymerized. Another important issue that needs to be solved is the role of septins, as they are considered part of the cytoskeleton.

The third category addresses the comparison between hyphal apical growth and septum formation. It is important to understand the differences or similarities between the apical and septal machineries. Is there a septal apparatus homologous to the tip polarisome that guides polymerization of the CAR and ensures tethering of vesicles that provide both plasma membrane and cell wall-synthesizing machinery to form the septa? The traffic of vesicles to support hyphal growth is unidirectional, but the vesicles directed to the septum may arrive to both sides of the septum from the immediate adjacent hyphal compartments (bidirectional traffic). Once the vesicles are fused to the septal membrane, is there an actual excess of membrane that needs to be recycled by endocytosis? The fusion of vesicles carrying the enzymes and precursors needed to form the septum wall may produce an excess of membrane and generate the need for plasma membrane remodeling via endocytosis (Delgado-Alvarez et al., 2010; Echauri-Espinosa et al., 2012). The presence of actin patches near septa associated with the ABP's characteristic of endocytic function (fimbrin, ARP-2/3 complex, and coronin) supports the involvement of endocytosis in septum development.

Finally, the fourth category entails the process of cell wall synthesis in the septum. It is important to understand why all CHSs participate in crosswall formation. Furthermore, we need more precise analyses to discern whether the composition of septal walls is indeed different than that of the lateral cell wall.

ACKNOWLEDGMENTS

Research in the Riquelme lab is supported by Mexican National Council for Science and Technology (CONACYT) grants U-45818Q, B0C022, and CONACYT-DFG 75306. Financial support for Mouriño Lab is obtained from CONACYT-DFG (75306) and CONACYT-Ciencia Básica (134771). We are indebted to E. Sánchez-León for producing Figures 3.2 and 3.5. We are grateful to D. Delgado-Alvarez, R. Fajardo-Somera, R. W. Roberson, S. Seiler, and L. Sánchez for sharing unpublished information. We thank S. Bartnicki-García for critically reading the manuscript and encouraging discussions. We are pleased to acknowledge the use of materials in our labs generated by P01 GM068087 "Functional Analysis of a Model Filamentous Fungus."

REFERENCES

Araujo-Palomares, C. L., Richthammer, C., Seiler, S., & Castro-Longoria, E. (2011). Functional characterization and cellular dynamics of the CDC-42–RAC–CDC-24 module in *Neurospora crassa*. *PLoS One*, *6*(11), e27148. http://dx.doi.org/10.1371/journal.pone.0027148.t001.

Bahler, J., Steever, A. B., Wheatley, S., Wang, Yl, Pringle, J. R., Gould, K. L., et al. (1998). Role of polo kinase and Mid1p in determining the site of cell division in fission yeast. *The Journal of Cell Biology*, *143*(6), 1603–1616.

Bardin, A. J., & Amon, A. (2001). MEN and SIN: What's the difference? *Nature Reviews. Molecular Cell Biology*, *2*(11), 815–826. http://dx.doi.org/10.1038/35099020.

Barrowman, J., Bhandari, D., Reinisch, K., & Ferro-Novick, S. (2006). TRAPP complexes in membrane traffic: Convergence through a common Rab. *Nature Reviews. Molecular Cell Biology*, *11*(11), 759–763.

Bartnicki-Garcia, S. (1987). Chitosomes and chitin biogenesis. *Food Hydrocolloids*, *1*(5–6), 353–358. http://dx.doi.org/10.1016/S0268-005X(87)80025-5.

Bartnicki-García, S. (1990). Role of vesicles in apical growth and a new mathematical model of hyphal morphogenesis. In I. B. Heath (Ed.), *Tip growth in plants and fungal cells* (pp. 211–232). San Diego: Academic Press.

Bartnicki-García, S. (2002). Hyphal tip growth: Outstanding questions. In H. D. Osiewacz (Ed.), *Molecular biology of fungal development* (pp. 29–58). New York: Marcel Dekker.

Bassilana, M., Hopkins, J., & Arkowitz, R. A. (2005). Regulation of the Cdc42/Cdc24 GTPase module during *Candida albicans* hyphal growth. *Eukaryotic Cell, 4*(3), 588–603. http://dx.doi.org/10.1128/EC.4.3.588-603.2005.

Berepiki, A., Lichius, A., Shoji, J. Y., Tilsner, J., & Read, N. D. (2010). F-actin dynamics in *Neurospora crassa*. *Eukaryotic Cell, 9*(4), 547–557. http://dx.doi.org/10.1128/EC.00253-09.

Bidlingmaier, S., Weiss, E. L., Seidel, C., Drubin, D. G., & Snyder, M. (2001). The Cbk1p pathway is important for polarized cell growth and cell separation in *Saccharomyces cerevisiae*. *Molecular and Cellular Biology, 21*(7), 2449–2462. http://dx.doi.org/10.1128/MCB.21.7.2449-2462.2001.

Bonifacino, J. S., & Glick, B. S. (2004). The mechanisms of vesicle budding and fusion. *Cell, 116*(2), 153–166.

Borkovich, K. A., Alex, L. A., Yarden, O., Freitag, M., Turner, G. E., Read, N. D., et al. (2004). Lessons from the genome sequence of *Neurospora crassa*: Tracing the path from genomic blueprint to multicellular organism. *Microbiology and Molecular Biology Reviews, 68*(1), 1–108.

Bowman, B. J., Draskovic, M., Freitag, M., & Bowman, E. J. (2009). Structure and distribution of organelles and cellular location of calcium transporters in *Neurospora crassa*. *Eukaryotic Cell, 8*(12), 1845–1855. http://dx.doi.org/10.1128/EC.00174-09.

Boyce, K. J., Hynes, M. J., & Andrianopoulos, A. (2005). The Ras and Rho GTPases genetically interact to co-ordinately regulate cell polarity during development in *Penicillium marneffei*. *Molecular Microbiology, 55*(5), 1487–1501. http://dx.doi.org/10.1111/j.1365-2958.2005.04485.x.

Bracker, C. E. (1967). Ultrastructure of fungi. *Annual Review of Phytopathology, 5*(1), 343–372. http://dx.doi.org/10.1146/annurev.py.05.090167.002015.

Bracker, C. E., Ruiz-Herrera, J., & Bartnicki-Garcia, S. (1976). Structure and transformation of chitin synthetase particles (chitosomes) during microfibril synthesis in vitro. *Proceedings of the National Academy of Sciences of the United States of America, 73*(12), 4570–4574.

Bruno, K. S., Morrell, J. L., Hamer, J. E., & Staiger, C. J. (2001). SEPH, a Cdc7p orthologue from *Aspergillus nidulans*, functions upstream of actin ring formation during cytokinesis. *Molecular Microbiology, 42*(1), 3–12.

Butt, T. M., & Heath, I. B. (1988). The changing distribution of actin and nuclear behavior during the cell cycle of the mite-pathogenic fungus *Neozygites* sp. *European Journal of Cell Biology, 46*(3), 499–505.

Cai, H., Reinisch, K., & Ferro-Novick, S. (2007). Coats, tethers, Rabs, and SNAREs work together to mediate the intracellular destination of a transport vesicle. *Developmental Cell, 12*(5), 671–682. http://dx.doi.org/10.1016/j.devcel.2007.04.005.

Calvert, M. E., Wright, G. D., Leong, F. Y., Chiam, K. H., Chen, Y., Jedd, G., et al. (2011). Myosin concentration underlies cell size-dependent scalability of actomyosin ring constriction. *The Journal of Cell Biology, 195*(5), 799–813. http://dx.doi.org/10.1083/jcb.201101055.

Chandler, D. E., & Roberson, R. W. (2008). *Bioimaging: Current techniques in light and electron microscopy*. Sudbury, MA: Jones and Bartlett Publishers.

Chang, F., Woollard, A., & Nurse, P. (1996). Isolation and characterization of fission yeast mutants defective in the assembly and placement of the contractile actin ring. *Journal of Cell Science, 109*(Pt. 1), 131–142.

Choquer, M., Boccara, M., Goncalves, I. R., Soulie, M. C., & Vidal-Cros, A. (2004). Survey of the *Botrytis cinerea* chitin synthase multigenic family through the analysis of six euascomycetes genomes. *European Journal of Biochemistry, 271*(11), 2153–2164. http://dx.doi.org/10.1111/j.1432-1033.2004.04135.x.

Clutterbuck, A. J. (1970). Synchronous nuclear division and septation in *Aspergillus nidulans*. *Journal of General Microbiology, 60*, 133–135.

Coffman, V. C., Nile, A. H., Lee, I. J., Liu, H., & Wu, J. Q. (2009). Roles of formin nodes and myosin motor activity in Mid1p-dependent contractile-ring assembly during fission yeast cytokinesis. *Molecular Biology of the Cell, 20*(24), 5195–5210. http://dx.doi.org/10.1091/mbc.E09-05-0428.

Collinge, A. J., & Markham, P. (1985). Woronin bodies rapidly plug septal pores of severed *Penicillium chrysogenum* hyphae. *Experimental Mycology, 9*(1), 80–85. http://dx.doi.org/10.1016/0147-5975(85)90051-9.

Colot, H. V., Park, G., Turner, G. E., Ringelberg, C., Crew, C. M., Litvinkova, L., et al. (2006). A high-throughput gene knockout procedure for *Neurospora* reveals functions for multiple transcription factors. *Proceedings of the National Academy of Sciences of the United States of America, 103*(27), 10352–10357. http://dx.doi.org/10.1073/pnas.0601456103.

Cortes, J. C. G., Ishiguro, J., Durán, A., & Ribas, J. C. (2002). Localization of the (1,3) beta-D-glucan synthase catalytic subunit homologue Bgs1p/Cps1p from fission yeast suggests that it is involved in septation, polarized growth, mating, spore wall formation and spore germination. *Journal of Cell Science, 115*(21), 4081–4096. http://dx.doi.org/10.1242/jcs.00085.

Das, M., Wiley, D. J., Chen, X., Shah, K., & Verde, F. (2009). The conserved NDR kinase Orb6 controls polarized cell growth by spatial regulation of the small GTPase Cdc42. *Current Biology, 19*(15), 1314–1319. http://dx.doi.org/10.1016/j.cub.2009.06.057.

Das, M., Wiley, D. J., Medina, S., Vincent, H. A., Larrea, M., Oriolo, A., et al. (2007). Regulation of cell diameter, For3p localization, and cell symmetry by fission yeast Rho-GAP Rga4p. *Molecular Biology of the Cell, 18*(6), 2090–2101. http://dx.doi.org/10.1091/mbc.E06-09-0883.

Delgado-Alvarez, D. L., Callejas-Negrete, O. A., Gomez, N., Freitag, M., Roberson, R. W., Smith, L. G., et al. (2010). Visualization of F-actin localization and dynamics with live cell markers in *Neurospora crassa*. *Fungal Genetics and Biology, 47*(7), 573–586. http://dx.doi.org/10.1016/j.fgb.2010.03.004.

Delgado-Alvarez, D. L., Seiler, S., Bartnicki-Garcia, S., & Mourino-Perez, R. (2013). Septum formation starts with the establishment of a septal actin tangle (SAT) at future septation sites. Paper presented at the 27th fungal genetics conference, Asilomar, Pacific Grove, California, USA.

DeMay, B. S., Meseroll, R. A., Occhipinti, P., & Gladfelter, A. S. (2009). Regulation of distinct septin rings in a single cell by Elm1p and Gin4p kinases. *Molecular Biology of the Cell, 20*(8), 2311–2326. http://dx.doi.org/10.1091/mbc.E08-12-1169.

Dettmann, A., Illgen, J., Marz, S., Schurg, T., Fleissner, A., & Seiler, S. (2012). The NDR kinase scaffold HYM1/MO25 is essential for MAK2 MAP kinase signaling in *Neurospora crassa*. *PLoS Genetics, 8*(9), e1002950. http://dx.doi.org/10.1371/journal.pgen.1002950.

Dijkgraaf, G. J., Abe, M., Ohya, Y., & Bussey, H. (2002). Mutations in Fks1p affect the cell wall content of beta-1,3- and beta-1,6-glucan in *Saccharomyces cerevisiae*. *Yeast, 19*(8), 671–690. http://dx.doi.org/10.1002/yea.866.

Douglas, C. M. (2001). Fungal B-(1,3)-D-glucan synthesis. *Medical Mycology, 39*, 55–66.

Dunkler, A., & Wendland, J. (2007). Candida albicans Rho-type GTPase-encoding genes required for polarized cell growth and cell separation. *Eukaryotic Cell, 6*(5), 844–854. http://dx.doi.org/10.1128/EC.00201-06.

Dynesen, J., & Nielsen, J. (2003). Branching is coordinated with mitosis in growing hyphae of Aspergillus nidulans. *Fungal Genetics and Biology*, *40*(1), 15–24. http://dx.doi.org/10.1016/S1087-1845(03)00053-7.

Echauri-Espinosa, R. O., Callejas-Negrete, O. A., Roberson, R. W., Bartnicki-Garcia, S., & Mourino-Perez, R. R. (2012). Coronin is a component of the endocytic collar of hyphae of *Neurospora crassa* and is necessary for normal growth and morphogenesis. *PLoS One*, *7*(5), e38237. http://dx.doi.org/10.1371/journal.pone.0038237.

Evangelista, M., Pruyne, D., Amberg, D. C., Boone, C., & Bretscher, A. (2002). Formins direct Arp2/3-independent actin filament assembly to polarize cell growth in yeast. *Nature Cell Biology*, *4*(1), 32–41. http://dx.doi.org/10.1038/ncb718.

Fajardo-Somera, R. A., Bowman, B., & Riquelme, M. (2013). The plasma membrane proton pump PMA-1 gets incorporated to distal parts of the hyphae independently of the Spitzenkörper in *Neurospora crassa*. *Eukaryotic Cell*, *12*(8). http://dx.doi.org/10.1128/EC.00328-12.

Fajardo-Somera, R. A., Roberson, R. W., Johnk, B., Bayram, O., Braus, G. H., & Riquelme, M. (2013). The role and traffic of chitin synthases in Neurospora crassa. Paper presented at the 27th fungal genetics conference.

Fiddy, C., & Trinci, A. P. J. (1976). Mitosis, septation, branching and the duplication cycle in *Aspergillus nidulans*. *Journal of General Microbiology*, *97*(2), 169–184.

Field, C. M., & Kellogg, D. (1999). Septins: Cytoskeletal polymers or signalling GTPases? *Trends in Cell Biology*, *9*(10), 387–394.

Fleissner, A., & Glass, N. L. (2007). SO, a protein involved in hyphal fusion in *Neurospora crassa*, localizes to septal plugs. *Eukaryotic Cell*, *6*(1), 84–94. http://dx.doi.org/10.1128/EC.00268-06.

Fleissner, A., Leeder, A. C., Roca, M. G., Read, N. D., & Glass, N. L. (2009). Oscillatory recruitment of signaling proteins to cell tips promotes coordinated behavior during cell fusion. *Proceedings of the National Academy of Sciences of the United States of America*, *106*(46), 19387–19392. http://dx.doi.org/10.1073/pnas.0907039106.

Fleissner, A., Sarkar, S., Jacobson, D. J., Roca, M. G., Read, N. D., & Glass, N. L. (2005). The so locus is required for vegetative cell fusion and postfertilization events in *Neurospora crassa*. *Eukaryotic Cell*, *4*(5), 920–930. http://dx.doi.org/10.1128/ec.4.5.920-930.2005.

Free, S. J. (2013). Fungal cell wall organization and biosynthesis. *Advances in Genetics*, *81*, 33–82. http://dx.doi.org/10.1016/b978-0-12-407677-8.00002-6.

Fukuda, K., Yamada, K., Deoka, K., Yamashita, S., Ohta, A., & Horiuchi, H. (2009). Class III chitin synthase ChsB of *Aspergillus nidulans* localizes at the sites of polarized cell wall synthesis and is required for conidial development. *Eukaryotic Cell*, *8*(7), 945–956. http://dx.doi.org/10.1128/EC.00326-08.

Galagan, J. E., Calvo, S. E., Borkovich, K. A., Selker, E. U., Read, N. D., Jaffe, D., et al. (2003). The genome sequence of the filamentous fungus *Neurospora crassa*. *Nature*, *422*(6934), 859–868. http://dx.doi.org/10.1038/nature01554.

Girbardt, M. (1979). A microfilamentous septal belt (FSB) during induction of cytokinesis in *Trametes versicolor* (L. ex Fr.). *Experimental Mycology*, *3*, 215–228.

Gladfelter, A. S. (2006). Nuclear anarchy: Asynchronous mitosis in multinucleated fungal hyphae. *Current Opinion in Microbiology*, *9*(6), 547–552. http://dx.doi.org/10.1016/j.mib.2006.09.002.

Gonzalez-Novo, A., Correa-Bordes, J., Labrador, L., Sanchez, M., Vazquez de Aldana, C. R., & Jimenez, J. (2008). Sep7 is essential to modify septin ring dynamics and inhibit cell separation during *Candida albicans* hyphal growth. *Molecular Biology of the Cell*, *19*(4), 1509–1518. http://dx.doi.org/10.1091/mbc.E07-09-0876.

Gordon, C. L., Khalaj, V., Ram, A. F., Archer, D. B., Brookman, J. L., Trinci, A. P., et al. (2000). Glucoamylase::green fluorescent protein fusions to monitor protein secretion in *Aspergillus niger*. *Microbiology*, *146*(Pt. 2), 415–426.

Gould, K. L., & Simanis, V. (1997). The control of septum formation in fission yeast. *Genes & Development*, *11*(22), 2939–2951. http://dx.doi.org/10.1101/gad.11.22.2939.

Grallert, A., Connolly, Y., Smith, D. L., Simanis, V., & Hagan, I. M. (2012). The S. pombe cytokinesis NDR kinase Sid2 activates Fin1 NIMA kinase to control mitotic commitment through Pom1/Wee1. *Nature Cell Biology*, *14*(7), 738–745. http://dx.doi.org/10.1038/ncb2514.

Gull, K. (1978). Form and function of septa in filamentous fungi. In J. E. Smith & D. R. Berrys (Eds.), *Developmental mycology* (pp. 78–93). New York: John Wiley and Sons.

Gupta, G. D., & Heath, I. B. (2000). A tip-high gradient of a putative plasma membrane SNARE approximates the exocytotic gradient in hyphal apices of the fungus *Neurospora crassa*. *Fungal Genetics and Biology*, *29*(3), 187–199. http://dx.doi.org/10.1006/fgbi.2000.1200.

Gupta, S., & McCollum, D. (2011). Crosstalk between NDR kinase pathways coordinates cell cycle dependent actin rearrangements. *Cell Division*, *6*, 19. http://dx.doi.org/10.1186/1747-1028-6-19.

Harris, S. D. (1997). The duplication cycle in *Aspergillus nidulans*. *Fungal Genetics and Biology*, *22*(1), 1–12. http://dx.doi.org/10.1006/fgbi.1997.0990.

Harris, S. D. (2001). Septum formation in *Aspergillus nidulans*. *Current Opinion in Microbiology*, *4*(6), 736–739.

Harris, S. D., Hamer, L., Sharpless, K. E., & Hamer, J. E. (1997). The *Aspergillus nidulans* sepA gene encodes an FH1/2 protein involved in cytokinesis and the maintenance of cellular polarity. *EMBO Journal*, *16*(12), 3474–3483. http://dx.doi.org/10.1093/emboj/16.12.3474.

Harris, E. S., Li, F., & Higgs, H. N. (2004). The mouse formin, FRLalpha, slows actin filament barbed end elongation, competes with capping protein, accelerates polymerization from monomers, and severs filaments. *Journal of Biological Chemistry*, *279*(19), 20076–20087. http://dx.doi.org/10.1074/jbc.M312718200.

Harris, S. D., & Momany, M. (2004). Polarity in filamentous fungi: Moving beyond the yeast paradigm. *Fungal Genetics and Biology*, *41*(4), 391–400.

Harris, S. D., Morrell, J. L., & Hamer, J. E. (1994). Identification and characterization of *Aspergillus nidulans* mutants defective in cytokinesis. *Genetics*, *136*(2), 517–532.

Hartwell, L. H. (1971). Genetic control of the cell division cycle in yeast. IV. Genes controlling bud emergence and cytokinesis. *Experimental Cell Research*, *69*(2), 265–276.

Hayakawa, Y., Ishikawa, E., Shoji, J. Y., Nakano, H., & Kitamoto, K. (2011). Septum-directed secretion in the filamentous fungus *Aspergillus oryzae*. *Molecular Microbiology*, *81*(1), 40–55. http://dx.doi.org/10.1111/j.1365-2958.2011.07700.x.

Hernandez-Rodriguez, Y., Hastings, S., & Momany, M. (2012). The septin AspB in *Aspergillus nidulans* forms bars and filaments and plays roles in growth emergence and conidiation. *Eukaryotic Cell*, *11*(3), 311–323. http://dx.doi.org/10.1128/ec.05164-11.

Higashida, C., Miyoshi, T., Fujita, A., Oceguera-Yanez, F., Monypenny, J., Andou, Y., et al. (2004). Actin polymerization-driven molecular movement of mDia1 in living cells. *Science*, *303*(5666), 2007–2010. http://dx.doi.org/10.1126/science.1093923.

Hoch, H. C., & Howard, R. J. (1980). Ultrastructure of freeze-substituted hyphae of the basidiomycete Laetisaria arvalis. *Protoplasma*, *103*(3), 281–297. http://dx.doi.org/10.1007/Bf01276274.

Horiuchi, H. (2009). Functional diversity of chitin synthases of *Aspergillus nidulans* in hyphal growth, conidiophore development and septum formation. *Medical Mycology*, *47*(Suppl. 1), S47–S52. http://dx.doi.org/10.1080/13693780802213332.

Howard, R. J. (1981). Ultrastructural analysis of hyphal tip cell growth in fungi: Spitzenkorper, cytoskeleton and endomembranes after freeze-substitution. *Journal of Cell Science*, *48*, 89–103.

Hsu, S. C., TerBush, D., Abraham, M., & Guo, W. (2004). The exocyst complex in polarized exocytosis. *International Review of Cytology, 233*, 243–265.

Huang, J., Huang, Y., Yu, H., Subramanian, D., Padmanabhan, A., Thadani, R., et al. (2012). Nonmedially assembled F-actin cables incorporate into the actomyosin ring in fission yeast. *The Journal of Cell Biology, 199*(5), 831–847. http://dx.doi.org/10.1083/jcb.201209044.

Hunsley, D., & Gooday, G. W. (1974). The structure and development of septa in *Neurospora crassa*. *Protoplasma, 82*(1), 125–146.

Ichinomiya, M., Yamada, E., Yamashita, S., Ohta, A., & Horiuchi, H. (2005). Class I and class II chitin synthases are involved in septum formation in the filamentous fungus *Aspergillus nidulans*. *Eukaryotic Cell, 4*(6), 1125–1136. http://dx.doi.org/10.1128/EC.4.6.1125-1136.2005.

Jedd, G., & Chua, N.-H. (2000). A new self-assembled peroxisomal vesicle required for efficient resealing of the plasma membrane. *Nature Cell Biology, 2*, 226–231.

Jedd, G., & Pieuchot, L. (2012). Multiple modes for gatekeeping at fungal cell-to-cell channels. *Molecular Microbiology, 86*(6), 1291–1294. http://dx.doi.org/10.1111/mmi.12074.

Justa-Schuch, D., Heilig, Y., Richthammer, C., & Seiler, S. (2010). Septum formation is regulated by the RHO4-specific exchange factors BUD3 and RGF3 and by the landmark protein BUD4 in *Neurospora crassa*. *Molecular Microbiology, 76*(1), 220–235. http://dx.doi.org/10.1111/j.1365-2958.2010.07093.x.

Kamasaki, T., Osumi, M., & Mabuchi, I. (2007). Three-dimensional arrangement of F-actin in the contractile ring of fission yeast. *The Journal of Cell Biology, 178*(5), 765–771. http://dx.doi.org/10.1083/jcb.200612018.

Kaminsky, S. G. W., & Hamer, J. E. (1998). hyp loci control cell pattern formation in the vegetative mycelium of Aspergillus nidulans. *Genetics, 148*, 669–680.

Kang, M. S., & Cabib, E. (1986). Regulation of fungal cell wall growth: A guanine nucleotide-binding, proteinaceous component required for activity of (1–3)-beta-D-glucan synthase. *Proceedings of the National Academy of Sciences of the United States of America, 83*(16), 5808–5812.

Kim, J. M., Lu, L., Shao, R., Chin, J., & Liu, B. (2006). Isolation of mutations that bypass the requirement of the septation initiation network for septum formation and conidiation in *Aspergillus nidulans*. *Genetics, 173*(2), 685–696. http://dx.doi.org/10.1534/genetics.105.054304.

Kim, J. M., Zeng, C. J., Nayak, T., Shao, R., Huang, A. C., Oakley, B. R., et al. (2009). Timely septation requires SNAD-dependent spindle pole body localization of the septation initiation network components in the filamentous fungus *Aspergillus nidulans*. *Molecular Biology of the Cell, 20*(12), 2874–2884. http://dx.doi.org/10.1091/mbc.E08-12-1177.

Kovar, D. R. (2006). Cell polarity: Formin on the move. *Current Biology, 16*(14), R535–R538. http://dx.doi.org/10.1016/j.cub.2006.06.039.

Kovar, D. R., & Pollard, T. D. (2004). Progressing actin: Formin as a processive elongation machine. *Nature Cell Biology, 6*(12), 1158–1159. http://dx.doi.org/10.1038/ncb1204-1158.

Kovar, D. R., Sirotkin, V., & Lord, M. (2011). Three's company: The fission yeast actin cytoskeleton. *Trends in Cell Biology, 21*(3), 177–187. http://dx.doi.org/10.1016/j.tcb.2010.11.001.

Krapp, A., & Simanis, V. (2005). Cell division: SIN, cytokinesis and ethanol dependency. *Current Biology, 15*(15), R605–R607. http://dx.doi.org/10.1016/j.cub.2005.07.045.

Kuratsu, M., Taura, A., Shoji, J. Y., Kikuchi, S., Arioka, M., & Kitamoto, K. (2007). Systematic analysis of SNARE localization in the filamentous fungus *Aspergillus oryzae*. *Fungal Genetics and Biology, 44*(12), 1310–1323. http://dx.doi.org/10.1016/j.fgb.2007.04.012.

Kwon, M. J., Arentshorst, M., Roos, E. D., van den Hondel, C. A., Meyer, V., & Ram, A. F. (2011). Functional characterization of Rho GTPases in *Aspergillus niger* uncovers conserved and diverged roles of Rho proteins within filamentous fungi. *Molecular Microbiology*, *79*(5), 1151–1167. http://dx.doi.org/10.1111/j.1365-2958.2010.07524.x.

Lai, J., Koh, C. H., Tjota, M., Pieuchot, L., Raman, V., Chandrababu, K. B., et al. (2012). Intrinsically disordered proteins aggregate at fungal cell-to-cell channels and regulate intercellular connectivity. *Proceedings of the National Academy of Sciences of the United States of America*, *109*(39), 15781–15786. http://dx.doi.org/10.1073/pnas.1207467109.

Latgé, J. P., & Calderone, R. (2006). The fungal cell wall. In U. Kues, & R. Fischer (Eds.), *The Mycota*: *Vol. 1*. (pp. 73–104). Berlin, Heidelberg: Springer-Verlag.

Lesage, G., Sdicu, A. M., Menard, P., Shapiro, B., Hussein, S., & Bussey, H. (2004). Analysis of b-1,3-glucan assembly in *Saccharomyces cerevisiae* using a synthetic interaction network and altered sensitivity to caspofungin. *Genetics*, *167*(1), 35–49.

Lichius, A., Yanez-Gutierrez, M. E., Read, N. D., & Castro-Longoria, E. (2012). Comparative live-cell imaging analyses of SPA-2, BUD-6 and BNI-1 in *Neurospora crassa* reveal novel features of the filamentous fungal polarisome. *PLoS One*, *7*(1), e30372. http://dx.doi.org/10.1371/journal.pone.0030372.

Lindsey, R., Cowden, S., Hernandez-Rodriguez, Y., & Momany, M. (2010). Septins AspA and AspC are important for normal development and limit the emergence of new growth foci in the multicellular fungus *Aspergillus nidulans*. *Eukaryotic Cell*, *9*(1), 155–163. http://dx.doi.org/10.1128/EC.00269-09.

Liu, B., & Morris, N. R. (2000). A spindle pole body-associated protein, SNAD, affects septation and conidiation in *Aspergillus nidulans*. *Molecular & General Genetics*, *263*(3), 375–387.

Liu, F., Ng, S. K., Lu, Y., Low, W., Lai, J., & Jedd, G. (2008). Making two organelles from one: Woronin body biogenesis by peroxisomal protein sorting. *The Journal of Cell Biology*, *180*(2), 325–339. http://dx.doi.org/10.1083/jcb.200705049.

Longtine, M. S., DeMarini, D. J., Valencik, M. L., Al-Awar, O. S., Fares, H., De Virgilio, C., et al. (1996). The septins: Roles in cytokinesis and other processes. *Current Opinion in Cell Biology*, *8*(1), 106–119.

Lord, M. (2010). Cytokinesis mechanisms in yeast. *Nature Education*, *3*(9), 53.

Maerz, S., Dettmann, A., Ziv, C., Liu, Y., Valerius, O., Yarden, O., et al. (2009). Two NDR kinase-MOB complexes function as distinct modules during septum formation and tip extension in Neurospora crassa. *Molecular Microbiology*, *74*(3), 707–723. http://dx.doi.org/10.1111/j.1365-2958.2009.06896.x.

Maerz, S., & Seiler, S. (2010). Tales of RAM and MOR: NDR kinase signaling in fungal morphogenesis. *Current Opinion in Microbiology*, *13*(6), 663–671. http://dx.doi.org/10.1016/j.mib.2010.08.010.

Maerz, S., Ziv, C., Vogt, N., Helmstaedt, K., Cohen, N., Gorovits, R., et al. (2008). The nuclear Dbf2-related kinase COT1 and the mitogen-activated protein kinases MAK1 and MAK2 genetically interact to regulate filamentous growth, hyphal fusion and sexual development in *Neurospora crassa*. *Genetics*, *179*(3), 1313–1325. http://dx.doi.org/10.1534/genetics.108.089425.

Mahadevan, P. R., & Tatum, E. L. (1967). Localization of structural polymers in the cell wall of *Neurospora crassa*. *The Journal of Cell Biology*, *35*(2), 295–302.

Mahs, A., Ischebeck, T., Heilig, Y., Stenzel, I., Seiler, S., & Heilmann, I. (2012). The essential phosphoinositide kinase MSS-4 is required for polar hyphal morphogenesis, localizing to sites of growth and cell fusion in *Neurospora crassa*. *PLoS One*, *7*(12), e51454. http://dx.doi.org/10.1371/journal.pone.0051454.t001.

Maruyama, J., Escano, C. S., & Kitamoto, K. (2010). AoSO protein accumulates at the septal pore in response to various stresses in the filamentous fungus *Aspergillus oryzae*. *Biochemical and Biophysical Research Communications*, *391*(1), 868–873. http://dx.doi.org/10.1016/J.Bbrc.2009.11.154.

Maruyama, J., Kikuchi, S., & Kitamoto, K. (2006). Differential distribution of the endoplasmic reticulum network as visualized by the BipA-EGFP fusion protein in hyphal compartments across the septum of the filamentous fungus, *Aspergillus oryzae*. *Fungal Genetics and Biology*, *43*(9), 642–654. http://dx.doi.org/10.1016/j.fgb.2005.11.007.

Masai, K., Maruyama, J., Nakajima, H., & Kitamoto, K. (2003). In vivo visualization of the distribution of a secretory protein in *Aspergillus oryzae* hyphae using the RntA-EGFP fusion protein. *Bioscience, Biotechnology, and Biochemistry*, *67*(2), 455–459.

Mazanka, E., Alexander, J., Yeh, B. J., Charoenpong, P., Lowery, D. M., Yaffe, M., et al. (2008). The NDR/LATS family kinase Cbk1 directly controls transcriptional asymmetry. *PLoS Biology*, *6*(8), e203. http://dx.doi.org/10.1371/journal.pbio.0060203.

McCollum, D., & Gould, K. L. (2001). Timing is everything: Regulation of mitotic exit and cytokinesis by the MEN and SIN. *Trends in Cell Biology*, *11*(2), 89–95.

Meitinger, F., Palani, S., & Pereira, G. (2012). The power of MEN in cytokinesis. *Cell Cycle*, *11*(2), 219–228. http://dx.doi.org/10.4161/cc.11.2.18857.

Minke, P. F., Lee, I. H., & Plamann, M. (1999). Microscopic analysis of *Neurospora* ropy mutants defective in nuclear distribution. *Fungal Genetics and Biology*, *28*(1), 55–67. http://dx.doi.org/10.1006/fgbi.1999.1160.

Momany, M., & Hamer, J. E. (1997a). The Aspergillus nidulans septin encoding gene, aspB, is essential for growth. *Fungal Genetics and Biology*, *21*(1), 92–100. http://dx.doi.org/10.1006/Fgbi.1997.0967.

Momany, M., & Hamer, J. E. (1997b). Relationship of actin, microtubules, and crosswall synthesis during septation in *Aspergillus nidulans*. *Cell Motility and the Cytoskeleton*, *38*(4), 373–384. http://dx.doi.org/10.1002/(SICI)1097-0169(1997)38, 4<373::AID-CM7>3.0.CO;2-4.

Momany, M., Richardson, E. A., Van Sickle, C., & Jedd, G. (2002). Mapping Woronin body position in *Aspergillus nidulans*. *Mycologia*, *94*(2), 260–266.

Momany, M., & Taylor, I. (2000). Landmarks in the early duplication cycles of *Aspergillus fumigatus* and *Aspergillus nidulans*: Polarity, germ tube emergence and septation. *Microbiology*, *146*, 3279–3284.

Momany, M., Zhao, J., Lindsey, R., & Westfall, P. J. (2001). Characterization of the *Aspergillus nidulans* septin (asp) gene family. *Genetics*, *157*(3), 969–977.

Morris, N. R. (1975). Mitotic mutants of Aspergillus nidulans. *Genetics Research*, *26*(03), 237–254. http://dx.doi.org/10.1017/S0016672300016049.

Moseley, J. B., Sagot, I., Manning, A. L., Xu, Y., Eck, M. J., Pellman, D., et al. (2004). A conserved mechanism for Bni1- and mDia1-induced actin assembly and dual regulation of Bni1 by Bud6 and profilin. *Molecular Biology of the Cell*, *15*(2), 896–907. http://dx.doi.org/10.1091/mbc.E03-08-0621.

Mouriño-Pérez, R. R. (2013). Septum development in filamentous ascomycetes. *Fungal Biology Reviews*, *27*(1), 1–9.

Nakano, K., Mutoh, T., Arai, R., & Mabuchi, I. (2003). The small GTPase Rho4 is involved in controlling cell morphology and septation in fission yeast. *Genes to Cells*, *8*(4), 357–370.

Nelson, B., Kurischko, C., Horecka, J., Mody, M., Nair, P., Pratt, L., et al. (2003). RAM: A conserved signaling network that regulates Ace2p transcriptional activity and polarized morphogenesis. *Molecular Biology of the Cell*, *14*(9), 3782–3803. http://dx.doi.org/10.1091/mbc.E03-01-0018.

Ng, S. K., Liu, F., Lai, J., Low, W., & Jedd, G. (2009). A tether for Woronin body inheritance is associated with evolutionary variation in organelle positioning. *PLoS Genetics*, *5*(6), e1000521. http://dx.doi.org/10.1371/journal.pgen.1000521.

Novick, P., Medkova, M., Dong, G., Hutagalung, A., Reinisch, K., & Grosshans, B. (2006). Interactions between Rabs, tethers, SNAREs and their regulators in exocytosis. *Biochemical Society Transactions, 34,* 683–686.

Pantazopoulou, A., & Peñalva, M. A. (2009). Organization and dynamics of the *Aspergillus nidulans* Golgi during apical extension and mitosis. *Molecular Biology of the Cell, 20*(20), 4335–4347. http://dx.doi.org/10.1091/mbc.E09-03-0254.

Park, H. O., & Bi, E. (2007). Central roles of small GTPases in the development of cell polarity in yeast and beyond. *Microbiology and Molecular Biology Reviews, 71*(1), 48–96. http://dx.doi.org/10.1128/MMBR.00028-06.

Pelham, H. R. B. (1999). SNAREs and the secretory pathway—Lessons from yeast. *Experimental Cell Research, 247,* 1–8.

Perez, P., & Rincon, S. A. (2010). Rho GTPases: Regulation of cell polarity and growth in yeasts. *Biochemical Journal, 426*(3), 243–253. http://dx.doi.org/10.1042/BJ20091823.

Pieuchot, L., & Jedd, G. (2012). Peroxisome assembly and functional diversity in eukaryotic microorganisms. *Annual Review of Microbiology, 66,* 237–263. http://dx.doi.org/10.1146/annurev-micro-092611-150126.

Plamann, M., Minke, P. E., Tinsley, J. H., & Bruno, K. S. (1994). Cytoplasmic dynein and actin-related protein Arp1 are required for normal nuclear distribution in filamentous fungi. *The Journal of Cell Biology, 127*(1), 139–149.

Pollard, T. D. (2008). Progress towards understanding the mechanism of cytokinesis in fission yeast. *Biochemical Society Transactions, 36*(Pt. 3), 425–430. http://dx.doi.org/10.1042/BST0360425.

Pruyne, D., Evangelista, M., Yang, C., Bi, E., Zigmond, S., Bretscher, A., et al. (2002). Role of formins in actin assembly: Nucleation and barbed-end association. *Science, 297*(5581), 612–615. http://dx.doi.org/10.1126/science.1072309.

Raju, N. B. (1992). Genetic control of the sexual cycle in *Neurospora*. *Mycological Research, 96,* 241–262.

Rasmussen, C. G., & Glass, N. L. (2005). A Rho-type GTPase, rho-4, is required for septation in *Neurospora crassa*. *Eukaryotic Cell, 4*(11), 1913–1925. http://dx.doi.org/10.1128/EC.4.11.1913-1925.2005.

Rasmussen, C. G., & Glass, N. L. (2007). Localization of RHO-4 indicates differential regulation of conidial versus vegetative septation in the filamentous fungus *Neurospora crassa*. *Eukaryotic Cell, 6*(7), 1097–1107. http://dx.doi.org/10.1128/EC.00050-07.

Rasmussen, C. G., Morgenstein, R. M., Peck, S., & Glass, N. L. (2008). Lack of the GTPase rho-4 in *Neurospora crassa* causes a reduction in numbers and aberrant stabilization of microtubules at hyphal tips. *Fungal Genetics and Biology, 45*(6), 1027–1039. http://dx.doi.org/10.1016/J.Fgb.2008.02.006.

Read, N. D. (2011). Exocytosis and growth do not occur only at hyphal tips. *Molecular Microbiology, 81*(1), 4–7. http://dx.doi.org/10.1111/j.1365-2958.2011.07702.x.

Reinhardt, M. O. (1892). Das Wachsthum der Pilzhyphen. *Jahrbucher fur wissenschaftliche Botanik, 23,* 479–566.

Richthammer, C., Enseleit, M., Sanchez-Leon, E., Marz, S., Heilig, Y., Riquelme, M., et al. (2012). RHO1 and RHO2 share partially overlapping functions in the regulation of cell wall integrity and hyphal polarity in *Neurospora crassa*. *Molecular Microbiology, 85*(4), 716–733. http://dx.doi.org/10.1111/j.1365-2958.2012.08133.x.

Riquelme, M., & Bartnicki-García, S. (2008). Advances in understanding hyphal morphogenesis: Ontogeny, phylogeny and cellular localization of chitin synthases. *Fungal Biology Reviews, 22*(2), 56–70. http://dx.doi.org/10.1016/j.fbr.2008.05.003.

Riquelme, M., Bartnicki-Garcia, S., González-Prieto, J. M., Sánchez-León, E., Verdín-Ramos, J. A., Beltrán-Aguilar, A., et al. (2007). Spitzenkörper localization and intracellular traffic of green fluorescent protein-labeled CHS-3 and CHS-6 chitin synthases in living hyphae of *Neurospora crassa*. *Eukaryotic Cell*, *6*(10), 1853–1864. http://dx.doi.org/10.1128/ec.00088-07.

Riquelme, M., Yarden, O., Bartnicki-Garcia, S., Bowman, B., Castro-Longoria, E., Free, S. J., et al. (2011). Architecture and development of the Neurospora crassa hypha—A model cell for polarized growth. *Fungal Biology*, *115*(6), 446–474. http://dx.doi.org/10.1016/j.funbio.2011.02.008.

Roberson, R. W. (1992). The actin cytoskeleton in hyphal cells of Sclerotium rolfsii. *Mycologia*, *84*(1), 41–51. http://dx.doi.org/10.2307/3760400.

Romero, S., Le Clainche, C., Didry, D., Egile, C., Pantaloni, D., & Carlier, M. F. (2004). Formin is a processive motor that requires profilin to accelerate actin assembly and associated ATP hydrolysis. *Cell*, *119*(3), 419–429. http://dx.doi.org/10.1016/j.cell.2004.09.039.

Rosenberg, R. F., & Kessel, M. (1967). Synchrony of nuclear replication in individual hyphae of *Aspergillus nidulans*. *Journal of Bacteriology*, *94*(5), 1464–1469.

Rosenberg, J. A., Tomlin, G. C., McDonald, W. H., Snydsman, B. E., Muller, E. G., Yates, J. R., 3rd., et al. (2006). Ppc89 links multiple proteins, including the septation initiation network, to the core of the fission yeast spindle-pole body. *Molecular Biology of the Cell*, *17*(9), 3793–3805. http://dx.doi.org/10.1091/mbc.E06-01-0039.

Rossman, K. L., Der, C. J., & Sondek, J. (2005). GEF means go: Turning on RHO GTPases with guanine nucleotide-exchange factors. *Nature Reviews. Molecular Cell Biology*, *6*(2), 167–180. http://dx.doi.org/10.1038/nrm1587.

Runeberg, P., Raudaskoski, M., & Virtanen, I. (1986). Cytoskeletal elements in the hyphae of the homobasidiomycete *Schizophyllum commune* visualized with indirect immunofluorescence and Nbd-Phallacidin. *European Journal of Cell Biology*, *41*(1), 25–32.

Sagot, I., Rodal, A. A., Moseley, J., Goode, B. L., & Pellman, D. (2002). An actin nucleation mechanism mediated by Bni1 and profilin. *Nature Cell Biology*, *4*(8), 626–631. http://dx.doi.org/10.1038/ncb834.

Salo, V., Niini, S. S., Virtanen, I., & Raudaskoski, M. (1989). Comparative immunocytochemistry of the cytoskeleton in filamentous fungi with dikaryotic and multinucieate hyphae. *Journal of Cell Science*, *94*, 11–24.

Sánchez, H. L., Freitag, M., Smith, K. M., & Riquelme, M. (2009). Role and dynamics of the septin hyp-1 in Neurospora crassa. Paper presented at the Xth international fungal biology conference, Ensenada, Baja California, Mexico.

Sánchez-León, E., Verdín, J., Freitag, M., Roberson, R. W., Bartnicki-Garcia, S., & Riquelme, M. (2011). Traffic of chitin synthase 1 (CHS-1) to the Spitzenkörper and developing septa in hyphae of *Neurospora crassa*: Actin dependence and evidence of distinct microvesicle populations. *Eukaryotic Cell*, *10*(5), 683–695. http://dx.doi.org/10.1128/ec.00280-10.

Sanders, S. L., & Herskowitz, I. (1996). The BUD4 protein of yeast, required for axial budding, is localized to the mother/BUD neck in a cell cycle-dependent manner. *The Journal of Cell Biology*, *134*(2), 413–427.

Schiel, J. A., & Prekeris, R. (2013). Membrane dynamics during cytokinesis. *Current Opinion in Cell Biology*, *25*(1), 92–98. http://dx.doi.org/10.1016/j.ceb.2012.10.012.

Schneper, L., Krauss, A., Miyamoto, R., Fang, S., & Broach, J. R. (2004). The Ras/protein kinase A pathway acts in parallel with the Mob2/Cbk1 pathway to effect cell cycle progression and proper bud site selection. *Eukaryotic Cell*, *3*(1), 108–120. http://dx.doi.org/10.1128/ec.3.1.108-120.2004.

Schurg, T., Brandt, U., Adis, C., & Fleissner, A. (2012). The *Saccharomyces cerevisiae* BEM1 homologue in *Neurospora crassa* promotes co-ordinated cell behaviour resulting in cell fusion. *Molecular Microbiology*, *86*(2), 349–366. http://dx.doi.org/10.1111/j.1365-2958.2012.08197.x.

Seiler, S., & Justa-Schuch, D. (2010). Conserved components, but distinct mechanisms for the placement and assembly of the cell division machinery in unicellular and filamentous ascomycetes. *Molecular Microbiology*, *78*(5), 1058–1076. http://dx.doi.org/10.1111/j.1365-2958.2010.07392.x.

Seiler, S., & Plamann, M. (2003). The genetic basis of cellular morphogenesis in the filamentous fungus *Neurospora crassa*. *Molecular Biology of the Cell*, *14*(11), 4352–4364. http://dx.doi.org/10.1091/mbc.E02-07-0433.

Seiler, S., Vogt, N., Ziv, C., Gorovits, R., & Yarden, O. (2006). The STE20/germinal center kinase POD6 interacts with the NDR kinase COT1 and is involved in polar tip extension in *Neurospora crassa*. *Molecular Biology of the Cell*, *17*(9), 4080–4092. http://dx.doi.org/10.1091/mbc.E06-01-0072.

Sharpless, K. E., & Harris, S. D. (2002). Functional characterization and localization of the *Aspergillus nidulans* formin SEPA. *Molecular Biology of the Cell*, *13*(2), 469–479. http://dx.doi.org/10.1091/mbc.01-07-0356.

Shoji, J. Y., Arioka, M., & Kitamoto, K. (2008). Dissecting cellular components of the secretory pathway in filamentous fungi: Insights into their application for protein production. *Biotechnology Letters*, *30*(1), 7–14. http://dx.doi.org/10.1007/s10529-007-9516-1.

Si, H., Justa-Schuch, D., Seiler, S., & Harris, S. D. (2010). Regulation of septum formation by the Bud3-Rho4 GTPase module in *Aspergillus nidulans*. *Genetics*, *185*(1), 165–176. http://dx.doi.org/10.1534/genetics.110.114165.

Sohrmann, M., Fankhauser, C., Brodbeck, C., & Simanis, V. (1996). The dmf1/mid1 gene is essential for correct positioning of the division septum in fission yeast. *Genes & Development*, *10*(21), 2707–2719.

Steele, G. C., & Trinci, A. P. J. (1977). Effect of temperature and temperature shifts on growth and branching of a wild type and a temperature sensitive colonial mutant (Cot 1) of *Neurospora crassa*. *Archives of Microbiology*, *113*(1–2), 43–48.

Taheri-Talesh, N., Xiong, Y., & Oakley, B. R. (2012). The functions of myosin II and myosin V homologs in tip growth and septation in *Aspergillus nidulans*. *PLoS One*, *7*(2), e31218. http://dx.doi.org/10.1371/journal.pone.0031218.

Takeshita, N., Ohta, A., & Horiuchi, H. (2005). CsmA, a class V chitin synthase with a myosin motor-like domain, is localized through direct interaction with the actin cytoskeleton in *Aspergillus nidulans*. *Molecular Biology of the Cell*, *16*(4), 1961–1970. http://dx.doi.org/10.1091/mbc.E04-09-0761.

Takeshita, N., Vienken, K., Rolbetzki, A., & Fischer, R. (2007). The *Aspergillus nidulans* putative kinase, KfsA (kinase for septation), plays a role in septation and is required for efficient asexual spore formation. *Fungal Genetics and Biology*, *44*(11), 1205–1214. http://dx.doi.org/10.1016/j.fgb.2007.03.006.

Tanabe, S., & Kamada, T. (1994). The role of astral microtubules in conjugate division in the dikaryon of *Coprinus cinereus*. *Experimental Mycology*, *18*(4), 338–348. http://dx.doi.org/10.1016/S0147-5975(06)80007-1.

Tenney, K., Hunt, I., Sweigard, J., Pounder, J. I., McClain, C., Bowman, E. J., et al. (2000). Hex-1, a gene unique to filamentous fungi, encodes the major protein of the Woronin body and functions as a plug for septal pores. *Fungal Genetics and Biology*, *31*(3), 205–217. http://dx.doi.org/10.1006/fgbi.2000.1230.

TerBush, D. R., Maurice, T., Roth, D., & Novick, P. (1996). The Exocyst is a multiprotein complex required for exocytosis in *Saccharomyces cerevisiae*. *EMBO Journal*, *15*(23), 6483–6494.

Tolliday, N., VerPlank, L., & Li, R. (2002). Rho1 directs formin-mediated actin ring assembly during budding yeast cytokinesis. *Current Biology*, *12*(21), 1864–1870.

Tomlin, G. C., Morrell, J. L., & Gould, K. L. (2002). The spindle pole body protein Cdc11p links Sid4p to the fission yeast septation initiation network. *Molecular Biology of the Cell*, *13*(4), 1203–1214. http://dx.doi.org/10.1091/mbc.01-09-0455.

Toyn, J. H., & Johnston, L. H. (1994). The Dbf2 and Dbf20 protein kinases of budding yeast are activated after the metaphase to anaphase cell cycle transition. *EMBO Journal*, *13*(5), 1103–1113.

Trinci, A. P. J. (1979). The duplication cycle and branching in fungi. In J. H. Burnett & A. P. J. Trinci (Eds.), *Fungal walls and hyphal growth* (p. 416). Cambridge: Cambridge University Press.

Trinci, A. P. J., & Collinge, A. J. (1973). Structure and plugging of septa of wild type and spreading colonial mutants of *Neurospora crassa*. *Archives of Microbiology, 91*(4), 355–364.

Trinci, A. P. J., & Morris, N. R. (1979). Morphology and growth of a temperature-sensitive mutant of *Aspergillus nidulans* which forms aseptate mycelia at non-permissive temperatures. *Journal of General Microbiology, 114*, 53–59.

Upadhyay, S., & Shaw, B. D. (2008). The role of actin, fimbrin and endocytosis in growth of hyphae in *Aspergillus nidulans*. *Molecular Microbiology, 68*(3), 690–705. http://dx.doi.org/10.1111/j.1365-2958.2008.06178.x.

Vavylonis, D., Wu, J. Q., Hao, S., O'Shaughnessy, B., & Pollard, T. D. (2008). Assembly mechanism of the contractile ring for cytokinesis by fission yeast. *Science, 319*(5859), 97–100. http://dx.doi.org/10.1126/science.1151086.

Verdín, J., Bartnicki-Garcia, S., & Riquelme, M. (2009). Functional stratification of the Spitzenkörper of *Neurospora crassa*. *Molecular Microbiology, 74*(5), 1044–1053. http://dx.doi.org/10.1111/j.1365-2958.2009.06917.x.

VerPlank, L., & Li, R. (2005). Cell cycle-regulated trafficking of Chs2 controls actomyosin ring stability during cytokinesis. *Molecular Biology of the Cell, 16*(5), 2529–2543. http://dx.doi.org/10.1091/mbc.E04-12-1090.

Versele, M., & Thorner, J. (2005). Some assembly required: Yeast septins provide the instruction manual. *Trends in Cell Biology, 15*(8), 414–424.

Vogt, N., & Seiler, S. (2008). The RHO1-specific GTPase-activating protein LRG1 regulates polar tip growth in parallel to Ndr kinase signaling in *Neurospora*. *Molecular Biology of the Cell, 19*(11), 4554–4569. http://dx.doi.org/10.1091/mbc.E07-12-1266.

Wang, A., Raniga, P. P., Lane, S., Lu, Y., & Liu, H. (2009). Hyphal chain formation in *Candida albicans*: Cdc28-Hgc1 phosphorylation of Efg1 represses cell separation genes. *Molecular and Cellular Biology, 29*(16), 4406–4416. http://dx.doi.org/10.1128/MCB.01502-08.

Wendland, J. (2003). Analysis of the landmark protein Bud3 of *Ashbya gossypii* reveals a novel role in septum construction. *EMBO Reports, 4*(2), 200–204. http://dx.doi.org/10.1038/sj.embor.embor727.

Wennerberg, K., Rossman, K. L., & Der, C. J. (2005). The Ras superfamily at a glance. *Journal of Cell Science, 118*(Pt. 5), 843–846. http://dx.doi.org/10.1242/jcs.01660.

Westfall, P. J., & Momany, M. (2002). *Aspergillus nidulans* septin aspB plays pre- and postmitotic roles in septum, branch, and conidiophore development. *Molecular Biology of the Cell, 13*, 110–118. http://dx.doi.org/10.1091/mbc.;01-06-0312.

Wolkow, T. D., Harris, S. D., & Hamer, J. E. (1996). Cytokinesis in *Aspergillus nidulans* is controlled by cell size, nuclear positioning and mitosis. *Journal of Cell Science, 109*, 2179–2188.

Wu, J. Q., Kuhn, J. R., Kovar, D. R., & Pollard, T. D. (2003). Spatial and temporal pathway for assembly and constriction of the contractile ring in fission yeast cytokinesis. *Developmental Cell, 5*(5), 723–734.

Wu, J. Q., Sirotkin, V., Kovar, D. R., Lord, M., Beltzner, C. C., Kuhn, J. R., et al. (2006). Assembly of the cytokinetic contractile ring from a broad band of nodes in fission yeast. *The Journal of Cell Biology, 174*(3), 391–402. http://dx.doi.org/10.1083/jcb.200602032.

Yarden, O., Plamann, M., Ebbole, D. J., & Yanofsky, C. (1992). cot-1, a gene required for hyphal elongation in Neurospora crassa, encodes a protein kinase. *EMBO Journal, 11*(6), 2159–2166.

Yoshida, S., Kono, K., Lowery, D. M., Bartolini, S., Yaffe, M. B., Ohya, Y., et al. (2006). Polo-like kinase Cdc5 controls the local activation of Rho1 to promote cytokinesis. *Science, 313*(5783), 108–111. http://dx.doi.org/10.1126/science.1126747.

Zigmond, S. H., Evangelista, M., Boone, C., Yang, C., Dar, A. C., Sicheri, F., et al. (2003). Formin leaky cap allows elongation in the presence of tight capping proteins. *Current Biology, 13*(20), 1820–1823.

INDEX

Note: Page numbers followed by "*f*" indicate figures, and "*t*" indicate tables.

A

ABC transporters, 21–23
ABPs. *See* Actin and actin-binding proteins
    (ABPs)
Actin and actin-binding proteins (ABPs)
  assemblies, 108
  cytochalasin A blocks, septum formation,
    107–108
  cytoskeleton components, localization,
    109*f*
  formin, 108
  Lifeact–GFP, actin reporter, 108
  myosin, 108
  "search, capture, pull, and release" model,
    110
  tropomyosin, 108
ALG-3/4 endo-siRNAs, sperm
    development
  genetic analyses, 46
  MAGO12 mutant strain,
    phenotypes, 46
  misregulation, 47
  mutator and Eri genes, 46
  temperature dependent sterility,
    46–47
Antiviral defense
  deep sequencing, 38
  natural viruses infection, 38
  RNAi response, initiation, 37–38
Argonaute protein
  ALG-3 and ALG-4, 46
  C. elegans, 11*t*
  CSR-1, 20–21
  interacting partners, 10
  MAGO mutant strains, 20
  NRDE-3, 34–35
  PIWI subfamily, 2, 47
  RDE-1, 14–18, 20
  RISC, binding partner, 9–10
  short RNAs and, 2
  slicer activity, 20–21

B

Biogenesis and molecular mechanisms,
    miRNA
  Argonaute proteins, 10, 11*t*
  cofactors and modulators, 13
  dicer processing, product, 9–10
  pri-miRNAs, 9–10
  RISC, 9–10
  RNA polymerase II-transcribed genes,
    9–10

C

Caenorhabditis elegans
  Argonaute protein (*see* Argonaute
    protein)
  deep sequencing, 2
  miRNA function (*see* miRNAs)
  RNAi and endogenous siRNAs, 13–47
  21U-RNAs (*see* 21U-RNAs)
CAR. *See* Contractile actomyosin ring
    (CAR)
Cell wall, biosynthesis
  composition and structure, 113
  secretory pathway and exocytosis
    α-amylase AmyB, secreted proteins,
      117–118
    chitosomes, 116
    CHS and GSC, 116
    macrovesicles, 116
    plasma membrane proteins, 117
    vesicle fusion, 116
  septum-directed vesicles, transportation,
    118–119
  synthesizing enzymes
    chitin synthesis, 113–114
    CHSs, 113–114, 114*f*, 115*f*
    FKS, 115
    GSC, 115
    Lifeact–GFP, coexpression, 115*f*
Chitin synthases (CHSs), 113–114
Chromatin modulators, 27–29

CHSs. *See* Chitin synthases (CHSs)
Contractile actomyosin ring (CAR), 104
CSR-1 22G-RNA
    antisense to protein-coding genes, 44
    biogenesis and function, 41*f*
    chromosomal abnormalities, 43
    germline function, 44–45
    histone locus, overexpression, 44
    isoforms, 44–45
    pathway, developmental roles, 43–45
    P-granule defects, 43
    pre-mRNA processing, 41–42
    U7 snRNA, 44
Cytochalasin A block, 107–108
Cytokinesis, 101–102
Cytoskeleton components, 109*f*

D

Dicer. *See also* RNAi, pathway
    26G-RNA biogenesis, dependence,
        35–36
    immunoprecipitates, 20
    mutants, 43
    proteins coimmunoprecipitation, 32
    and RDE-4, 14–16
    -RRF-3 core complex, 32
Down syndrome cell adhesion molecule
    (Dscam), 75
Drosophila
    antifungal defense, 72
    immune responses
        AMPs, 73
        blood cells, 73–74
        Dscam, 75
        mediators and effector, 73
        melanization, 74–75
        phagocytosis, 75
        quinones, 74–75
    pattern recognition hypothesis, 72
    TLR4, 72
    toll signaling (*see* Toll signaling)
Dscam. *See* Down syndrome cell adhesion
    molecule (Dscam)
dsRNA-induced gene silencing, 24–27

E

Endochitinase, 101–102
Endogenous genes, 40–42

Endo-siRNA
    ADARs and endogenous RNAi, 36–37
    ALG-3/4 Argonautes, 32–34
    amplification and additional Argonautes,
        18–21
    discovery, 29–30
    ERGO-1 targets, 35
    26G- and 22G-, ERI pathway, 32–36, 33*f*
    GFP-sensor, silencing, 34–35
    HENN-1 RNA methyltransferase, 32–34
    mutator proteins, 21
    NRDE-3, somatic Argonaute protein,
        34–35
    RDE-1 and RDE-4, 36
    WAGO- and CSR-1-associated, 30–32
ERIC. *See* ERI-Dicer-1 complex (ERIC)
ERI-Dicer-1 complex (ERIC), 32

F

Formins, actin polymerization proteins
    BNI-1, 111
    conditional mutants, 111
    FH2 domain, 111
    SepA, 111

G

GEFs. *See* Guanine nucleotide exchange
        factors (GEFs)
Glucanase, 101–102
Glucan synthase complex (GSC), 115
22G-RNAs, 30–31. *See also* RNAi, pathway
26G-RNAs
    definition, 32–34
    Dicer dependence, 35–36
    sequences, 35–36
GSC. *See* Glucan synthase complex (GSC)
Guanine nucleotide exchange factors
        (GEFs), 105–107
Gut immunity and IMD signaling, 85–87

H

Hippo signaling, 85–87
Host-pathogen interaction, 87
Hunchback-like (HBL-1), transcription
        factor, 4

I

Immune deficiency (IMD) pathway
    gut immunity and signaling

enterocytes, characteristic, 85–87
Hippo signaling, 85–87
intestinal stem cells, 85–87
PGRP-LE, 85–87
host-pathogen interaction, 87
recognition and intracellular signaling
E3-ligase, 84
negative regulation, 85
PGRP-LCx, 83–84
Phi31, 83–84
receptor-adaptor complex, 84
TCT, 83–84
signaling pathway, 86f
Immunity, metabolism sensor, 87–88
Innate immunity
adaptive receptors, 72
IMD signaling pathway, 86f

L

Lifeact-GFP, actin reporter, 108

M

MAGO. See Multiple Argonaute (MAGO)
Melanization, 74–75
MEN. See Mitotic exit network (MEN)
miRNAs
biogenesis and molecular mechanisms,
9–13
developmental timing
cell divisions and cell-fate decisions, 3–4
lin-14 gene, 3–4
mutant screens, 3–4
mutations, 4
embryonic development
alg-1 and alg-2, embryonic lethality, 4
genetic screen, 5–6
lin-4 and let-7, 5–6
longevity and stress response
DAF-16-dependent life span
extension, 8
mir-71 function, 9
temperature-dependent mutant allele, 8
physiology
DNA damage-induced cell death, 7
mir-240/786 cluster, 8
muscle tissue, expression, 7
neuromuscular junctions, synaptic
transmission, 7

postembryonic development
defects, 7
EGF/RAS and Notch signaling
pathways, 6
loss-of-function studies, 6–7
mir-61 transcription, 6–7
nervous system, timing, 6
Mitotic exit network (MEN), 103–104
ModSP. See Modular serine protease
(ModSP)
Modular serine protease (ModSP), 78–80
MOR. See Morphogenesis-related NDR
network (MOR)
Morphogenesis
cell, 107
hyphal, 106–107
Morphogenesis-related NDR network
(MOR), 104
Multiple Argonaute (MAGO), 20
Mutator proteins, 21
Myosin, 108

N

Neurospora crassa
genome sequence, 101
septum ontogeny (see Septation)
septum wall biosynthesis (see Cell wall,
biosynthesis)

P

Peptidoglycan recognition protein-SA
(PGRP-SA), 76–77
PGRP-SA. See Peptidoglycan recognition
protein-SA (PGRP-SA)
Phagocytosis, 75
Primary miRNA transcripts (pri-miRNAs),
9–10
pri-miRNAs. See Primary miRNA
transcripts (pri-miRNAs)

Q

Quinones, 74–75

R

Rb cooperation, 45–46
RdRP. See RNA-dependent RNA
polymerases (RdRP)

Recognition and intracellular signaling, 83–85
Rho GTPases and GEFs, septation landmarks
  AgBud3, 106
  AnBud4, 106
  Bud4, 105–106
  BUD-6, 106–107
  cell morphogenesis, 107
  cortical rings, 105–106
  RHO-4, BUD-3, and BUD-4, 105
RISC. See RNA-induced- silencing complex (RISC)
RNA-dependent RNA polymerases (RdRP)
  amplification, 30
  EGO-1, 42, 43
  genes, 18–20
  module components, 30–31
  RNAi pathway, downstream components, 18
  role, 13–14
  RRF-3, 20
RNA-induced- silencing complex (RISC), 9–10
RNAi, pathway
  ABC transporters
    HAF-6, 21–22
    HAF-6 localization, 22–23
    -resistant mutants, 22–23
    second-site noncomplementation interaction, 22–23
    -silencing process, 21–22
  antisense siRNAs, target-dependent accumulation, 19f
  biological functions
    ALG-3/4 endo-siRNAs, sperm development, 46–47
    antiviral defense, 37–38
    CSR-1 22G-RNA pathway, developmental roles, 43–45
    endogenous genes, global effects, 40–42
    environmental adaptation, 42–43
    and Rb cooperation, 45–46
    transposons and repetitive elements, silencing, 38–40
  chromatin modulators

  genetic relationship, 28
  H3K36 methyltransferase MES-4, 27
  hypothetical genetic pathway, 28f
  Rb homolog lin-35, 27–28
  SynMuv suppressors, 28
  dsRNA-induced gene silencing, inheritance
    F1 generation, 25
    F4 generation, 25–26
    genomic analysis, 26
    HRDE-1/WAGO-9, 26–27
    NRDE-3 Argonaute, 24–25
    sense and antisense siRNAs, 26
    target-dependent siRNA amplification, 24–25
  endogenous, genomics and molecular features (see Endo-siRNA)
  hairpin precursors, 17–18
  mutator proteins
    mut-7 Au2 and rde-2/mut-8 genes, 21
    MUT-7 protein, 21
    transposon mobilization, 21
  RDE-4 and Dicer, 14–16
  RDE-1 Argonaute protein, 16–18
  RDE-10/RDE-11 complex, 23
  schematic, 15f
  siRNA amplification and additional Argonautes, 18–21
  systemic features
    channel SID-1, dsRNA import, 54–56, 55f
    SID-2, 56
    SID-3, 56
    SID-5, 57
  transcriptional silencing
    dsRNA-induced silencing, 23–24
    H3K9 methylation, 23–24
    NRDE-3, Argonaute protein, 23–24

S
Septal pores-associated proteins (SPA proteins), 119–120
Septation
  cytoskeletal machinery
    actin and actin-binding proteins, 107–110
    formins, actin polymerization proteins, 111

septins, 111–112
hyphal development
  cytokinesis, 101–102
  endochitinase, 101–102
  glucanase, 101–102
  nuclei, parasynchronous mitotic
    pattern, 102
  septum-degrading enzymes, 101–102
  solophenyl flavine, vital dye, 102–103
regulators and positional markers
  Rho GTPases and GEFs, landmarks,
    105–107
  SIN, 103–105
Septation initiation network (SIN)
  contractile actomyosin ring, 104
  Dbf2 subfamily, 104
  germinal center (GC) kinases, 104
  mitotic exit network and, 103–104
  MOB-NDR, 104
  NDR kinases, 104
  negative regulators, 104–105
  protein kinases, types, 104
Septins
  AspA and AspC, 112
  AspB, 112
  CDC-3, 111–112
  expression, 112
  identical localization patterns, 112
  multiseptin complexes, 111–112
  proteins, recruiting and organizing,
    111–112
Septum-directed vesicles, transportation
  BipA-EGFP, ER chaperone, 118–119
  exocytosis, 119
  SNAREs, vascular transport and active
    sorting, 118
  vesicle biogenesis and traffic, 118
  vesicle docking, 119
SIN. See Septation initiation network (SIN)
Solophenyl flavine, vital dye, 102–103
Spätzle-processing enzyme (SPE), 78–80
SPE. See Spätzle-processing enzyme (SPE)

T

TCT. See Tracheal cytotoxin (TCT)
TLR4. See Toll-like receptor 4 (TLR4)
Toll-like receptor 4 (TLR4), 72
Toll signaling

bacterial and fungal cell wall components,
    sensing
  cell wall, PGRP-SA binding, 78
  germ-line-coded receptors, 78
  GNBP3, binding affinity, 78
  Spz cleavage, 78
host-pathogen interaction
  nitric oxide, 83
  systemic immunity, 83
  WTA and, 82
intracellular signaling
  MyD88 recruitment, 81–82
  receptor-adaptor complex, 81–82
  Tube and Pelle, 81–82
necrotic-persephone axis, 75–76
PGRPs and GNBPs
  Osiris (osi) mutation, 76–77
  P-element-mediated partial deletion,
    77
  PGRP-SA, 76–77
  Semmelweis, 76–77
signaling pathway, 79f
Spz, ligand activation
  ModSP, 78–80
  paracrine role, 80–81
  signaling role, 80
  SPE, 78–80
TPM-1. See Tropomyosin (TPM-1)
Tracheal cytotoxin (TCT), 83–84
Transcriptional silencing, 23–24
Transposons and repetitive elements,
    silencing
  genome surveillance system, 38–40
  repetitive regions and foreign genomic
    elements, 39f
  RNAi-deficient mutants, 38–40
  WAGO-22G-RNAs, 38–40, 39f
Tropomyosin (TPM-1), 108

U

21U-RNAs (pi RNA)
  biogenesis
    autonomous expression, 49–50
    CTGTTTCA requirement, 49–50
    Forkhead protein-encoding genes,
      49–50
    genomic distributions, 47–49
    precursors, 50

21U-RNAs (pi RNA) (*Continued*)
  prg-1-dependent, 50
  production, 47–49
  YRNT motif, 48*f*
  biological functions
    computational prediction, 53
    endogenous gene regulation, 53
    fertility function, 51
    22G-RNA production and stable
      foreign DNA silencing, 51–54

Polycomb and Trithorax complex,
  52–53
transgene-silencing, 52–54
sterility phenotypes, 47

W
WAGO-22G-RNAs, 30–31.
  See also RNAi, pathway
WBs. *See* Woronin bodies (WBs)
Woronin bodies (WBs), 120